COUNCIL ON FOREIGN RELATIONS
LIBRARY

FEB 1 3 1992

58 EAST 68 STREET
NEW YORK, N.Y 10021

BUILDUP

BUILDUP

The Politics of Defense in the Reagan Era

DANIEL WIRLS

Cornell University Press

ITHACA AND LONDON

Copyright © 1992 by Cornell University

All rights reserved. Except for brief quotations in a review, this book, or parts thereof, must not be reproduced in any form without permission in writing from the publisher. For information, address Cornell University Press, 124 Roberts Place, Ithaca, New York 14850.

First published 1992 by Cornell University Press.

International Standard Book Number 0-8014-2442-9
Library of Congress Catalog Card Number 91-55075

Printed in the United States of America

Librarians: Library of Congress cataloging information appears on the last page of the book.

⊗ The paper in this book meets the minimum requirements of the American National Standard for Information Sciences—Permanence of Paper for Printed Library Materials, ANSI Z39.48-1984.

Contents

	List of Tables	vii
	Acknowledgments	ix
	Introduction Defense Policy and Domestic Politics	1
1	Continuity and Change in Cold War Politics	11
2	Military Policy and the Reagan Revolution	31
3	The Rise of the Nuclear Weapons Peace Movement	56
4	The Origins of Military Reform	79
5	The Nuclear Freeze Campaign and the Democratic Party	102
6	The Strategic Defense Initiative	133
7	Congressional Procurement Reform	169
8	From Buildup to Build-down	198
	Bibliography	227
	Index	241

Tables

1 Informed opinion on SALT II 26
2 Public opinion on defense spending, 1974–81 27
3 Changes in voting for president, 1976 and 1980 30
4 Military spending, fiscal years 1980–86 36
5 Growth of the capital-intensive defense budget 45
6 Sales and income for the LTV Corporation, 1981–85 49
7 Ideological diversity of the Military Reform Caucus compared to the Ninety-ninth Congress 93
8 Qualified support for a freeze 107
9 Scale of support for freeze movement objectives 107
10 Support for matching Soviet military power 108
11 SDI funding, 1984–89 155
12 Public priorities for government spending, October 1989 221

Acknowledgments

The exacerbation of cold war tensions and the military buildup in the late 1970s and early 1980s kindled in many people, I among them, a political and intellectual interest in nuclear weapons, arms control, and American defense policy. Not long into my graduate studies, I decided to specialize in American politics. Students with interests in defense policy tended to choose international relations, as most work on defense policy has been a product of the experts and epistemology of that field within political science. I chose instead to concentrate on American politics without abandoning my research interests in defense policy. This book, then, is one attempt to elucidate the relationship between American domestic politics and military policy.

I encountered nothing but support and encouragement, regardless of my personal intellectual headaches and misgivings, and I am grateful. First and foremost, I express my appreciation to Benjamin Ginsberg, Ned Lebow, Theodore Lowi, and Martin Shefter of Cornell University. Each made singular and significant contributions to this book, as well as to my broader education in politics and government. Martin Shefter, more than anyone, helped me to clarify the central ideas of my project. Ben Ginsberg contributed pivotal intellectual guidance, from which I will continue to benefit for years to come. I owe a significant debt to the Council for a Livable World and its former president, Jerome Grossman. My work there between

1982 and 1984 provided a wealth of experience with the peace movement and the politics of the defense policy process, as well as a substantial collection of peace movement literature. My book also profited from conversations with, and criticism from, many colleagues, including Michael K. Brown, Timothy Byrnes, John Garofano, Richard Garwin, Peter Katzenstein, Katherine Magraw, Diarmuid McGuire, Wendy Mink, Jonas Pontusson, Judith Reppy, Diane Schmidt, and Herman Schwartz. Alice Talbot eliminated many inconsistencies and awkward constructions. At Cornell University Press, Holly Bailey saw promise in an unrefined manuscript and had confidence in my ability to improve it. Janet Mais and Patty Peltekos provided acute editorial judgment. Readings of the manuscript by two anonymous reviewers provided both encouragement and important guidance. I single out Lynn Eden, whose insightful and exhaustive (as well as exhausting) commentaries on two versions of the manuscript are responsible for much of the improvement in form and substance.

Although my research sources were mostly primary and secondary written ones, I used interviews and written communications with about twenty government officials and political activists to fill in gaps and clarify factual uncertainties, especially regarding Star Wars and military reform. I thank all these people for their time and trouble. Several congressional offices were helpful and prompt in providing various documents, letters, and other information. The fine coverage of defense issues and Congress by the *Congressional Quarterly Weekly Report* and the *National Journal* alleviated some of my work. Some of the survey data I used was made available through the Inter-University Consortium for Political and Social Research.

Portions of Chapters 4 and 7 appeared as "Congress and the Politics of Military Reform," in *Armed Forces and Society* 17 (Summer 1991): 487–512. I thank the editors of the journal for allowing me to rework this material.

My gratitude extends to the Cornell Government Department and to Cornell University for their generous financial help, including a Sage Fellowship and summer support. The Peace Studies Program at Cornell provided a fellowship funded by the MacArthur Foundation, and the Cornell-in-Washington Program became a pleasant home away from home during a year of research in Washington, D.C. The University of California at Santa Cruz provided timely support in the form of a Regents' Junior Faculty Development Grant

Acknowledgments

and funding for research assistance from the university's Committee on Research. I am grateful to Steven Seto and Jason Kassel for their diligent work as research assistants while they were finishing their undergraduate degrees at Santa Cruz and to André Lambelet for his preparation of the index. Finally, I thank my colleagues in the Board of Politics at Santa Cruz, who found my work interesting enough to ask me to join their ranks.

DANIEL WIRLS

Santa Cruz, California

BUILDUP

Introduction

Defense Policy and Domestic Politics

The 1980s ushered in the most significant changes in American politics since the New Deal and ended with the most dramatic transformations in the international arena since World War II. The two-term Reagan presidency (1981–89) not only initiated and sustained the largest peacetime military buildup in American history but also altered traditional partisan alignments and revised the policy agenda of the welfare state. National security policy played a significant role in the effort by Reagan and the Republicans to forge a majority coalition and in the responses to that effort by liberal activists, the Democratic party, and Congress. The coalitions and institutions arrayed for and against the Reagan revolution used their defense policy agendas in their efforts to build or enlarge their constituencies and political influence. While they waged a genuine battle over the size and purpose of the military, they also fought a war for political power in which their principles and programs for national security were important weapons. Specifically, the politics of national security in the Reagan era derived from the conflicts among the three prominent and powerful defense policy agendas: those of the new cold war military buildup, the nuclear weapons peace movement, and military reform. Each agenda was linked closely to an important domestic coalition or institution—the arms buildup to the Reagan presidency and the Republicans, the peace movement to a coalition of liberal causes with significant influence in the Democratic party, and military reform to a bipartisan coalition in Congress. As a result, the conflicts had profound consequences not only for defense policy but for American government and public policy as well.

The interaction of national security policy with domestic politics has a rich history reaching back to the beginning of the Republic. For example, the Alien and Sedition Acts of 1798 were far more an attempt to stifle Republican opposition to Federalist policy than a prophylactic against pernicious foreign influence. Likewise, the War of 1812 had its origins in Republican desires to break the Federalist party as well as in Britain's Orders in Council. Other instances span the somewhat mythical era of nineteenth- and early twentieth-century isolationism to the internationalism of the cold war decades. Until World War II, however, the United States did not sustain a large permanent armed force. Consequently, while foreign policy played an ongoing role in United States politics, defense policy, or purely military preparations, only intermittently did so. After World War II, defense policy assumed continual importance because of the large military establishment equipped with nuclear weapons and in a permanent state of readiness. As we shall see in Chapter 1, the cold war was often fought at home as well as abroad.

In the 1980s, the defense debate and policy-making process moved to center stage. The candidacy of Ronald Reagan mobilized a revitalized Republican coalition, largely around the twin themes of military and economic decline under liberal foreign and social policies in the 1970s. Arguably the most clear and consistent Republican campaign promise was a rapid and sustained military buildup. Reagan tried to construct a new cold war consensus, in the wake of the Vietnam War and détente, by linking his diverse coalition to the popular themes of peace through strength and the the Soviet threat. Reagan's November 1980 victory ignited widespread speculation that a political realignment was in the making and with it, the end of the New Deal system of political alignments and policies. Reagan had won many voters to his cause, including a significant number who—at least until then—had identified themselves as Democrats and who were part of traditionally Democratic regions and constituencies.

Once he was in office, his administration commenced a restructuring of national policy through a combination of a rapid increase in defense spending, cuts in domestic programs, and significant tax cuts. The massive military buildup became the political fulcrum of Reaganomics. High defense spending coupled with tax cuts created record deficits. The Democrats found themselves with few if any popular programs left to promote; for having lost political control of the warfare state, they had lost the initiative in welfare and domestic policy as well. A new, higher plateau of military spending, growing deficits, and a bipartisan fear of new taxes constrained the political options of Democrats and aggravated political divisions in the party. The Reagan administration may ultimately have been unable to spend as much on the military or to cut domestic programs or taxes as much as it wanted, but in altering the fiscal equation and the policy priorities of the federal government it transformed the American political landscape.

Within a year after Reagan's inauguration, a burgeoning nuclear peace movement was emerging as the most serious threat to his political power and his agenda. Through its nuclear freeze campaign, the peace movement was evoking public anxiety about the nuclear arms race and nuclear war and building a massive and well-organized grass-roots movement. Reagan's own programs and rhetoric had done much to foster the widespread concern. The freeze proposal and shifting public opinion began to threaten the Washington consensus on a get-tough approach to the Soviets backed by a defense buildup centered on a vast nuclear weapons rearmament program. But the peace movement represented more than just anxiety about nuclear arms. Its seemingly infectious popularity made it a rallying cause for a general liberal counterattack on the Reagan revolution. The peace movement forged links of mutual interest with other constituencies and causes, including the women's, consumer, and environmental movements, blacks, elements of the labor movement, and Catholics. With these allies, it became one of the most important forces in the Democratic party, wielding considerable influence over the party's response to the president's programs. The central achievement of the freeze campaign was its transformation of the generalized anxiety about superpower relations into a focused debate on the ends of security policy. The issue became whether nuclear weapons in fact threatened our security rather than defended it. Ultimately, the peace movement could take credit for much of the pressure that eventually compelled Reagan to negotiate arms control with the Soviets, but it was in many ways a Pyrrhic victory. The movement failed to translate its success and popularity into enduring political power. It could forge only transient and fragile majorities in Congress, and it failed to elect Walter Mondale in 1984. The Democratic party, which had endorsed the freeze and made common cause with the peace movement, emerged from the 1984 election in a state of political disarray that could not be easily remedied.

In reaction to the contentious and politically treacherous struggle between the Reagan administration and the freeze movement and its allies, Congress turned to military reform. Its advocates championed the reformation of conventional war strategy, conventional weaponry, military organization, and the arms procurement process, with a growing emphasis on the eradication of waste, fraud, and abuse in the Pentagon and the arms industry. This national security agenda had the politically appealing potential to convert the struggle between the peace movement and the cold war coalition into one over the means rather than the ends of military policy. Military reform transformed many of the ideologically contentious and divisive issues of military policy into consensual and ethical issues about which there could be little disagreement. Consequently, a sizable and influential bloc of representatives and senators adopted the rhetoric and program of defense reform to balance conflicting forces for arms control, rearmament, and fiscal responsibility. With narrow majorities in both the Democratic House and the Republican Senate, Congress was unusually fragmented. Military reform provided Congress as an institution with some coherence on military issues, and individual members with a politically marketable position on national security. The congressional military reform coalition was able to find the circumstances and resources to manufacture a powerful political agenda around issues for which, initially, there had been no popular constituency. One example is weapons procurement reform, which allowed reformers to embrace defense spending limits *and* a stronger military with the same program. Although appeals for reform are nothing new, this one eclipsed all others in influence and institutional focus. In fact, each of the three defense policy movements of this period surpassed its predecessors in size, power, and political endurance. The scope and intensity of this tripartite conflict embodied an unparalleled domestication of national security policy.

The conflict over defense policy diminished toward the end of the Reagan presidency. Indeed, the election of 1988 was remarkable for the lack of debate, fervor, and vision on this most important and expensive subject. Nevertheless, the conflict left its political legacy. The following year brought sudden and unexpected events that fractured the long-standing framework of the American debate on defense policy but not the primacy of domestic politics. The transition from the buildup of the 1980s to the "post–cold war" retrenchment and restructuring of military policy as the new decade began would reinforce just how important domestic politics is to defense policy and vice versa.

This analysis stands in contrast to the dominant explanations of the causes and ramifications of U.S. national security policy. Political analysts generally recognize that military policies have origins in and consequences for domestic as well as international politics. As Samuel Huntington has pointed out, "Military policy . . . exists in two worlds. One is international politics, the world of the balance of power, wars and alliances, the subtle and the brutal uses of force and diplomacy to influence the behavior of other states. . . . The other world is domestic politics, the world of interest groups, political parties, social classes, with their conflicting interests and goals. . . . Any major decision in military policy influences and is influenced by both worlds."[1] Although Huntington and others have recognized this interactive relationship, social scientists and historians have tended to examine only certain of these interactions and influences to the near exclusion of others. The main reason for this is the traditional division of labor between students of international relations and students of American politics. Analysis of American defense policy and national security has been predominantly the domain of specialists in international security and relations; Americanists have paid the subject little attention. Consequently, certain domestic components and consequences of defense policy have been relatively neglected—to the detriment of insights into both American politics and international relations. Specifically, there have been few attempts to account for politicians' use of defense policy to shape domestic political competition; that is, politicians' use of defense policy to adjust to and influence domestic politics. Here domestic political competition refers to the ongoing contest for political power among public officials and their opponents, with an

---

1. Samuel P. Huntington, *The Common Defense* (New York, 1961), p. 1. Of course the observation that domestic politics influence *foreign policies*, and that politicians sometimes use foreign policy to cope with domestic problems, goes back to Thucydides and his *History of the Peloponnesian Wars*. Like Thucydides, modern commentators usually have focused on foreign policies and crises, not military policies specifically. Much like Huntington, Richard Aliano states that the "formulation (or more correctly, the re-formulation) of national security policy is to be found at the intersection of the international and domestic systems." He warns against adopting unadulterated versions of either international or domestic explanations. Aliano, *American Defense Policy from Eisenhower to Kennedy* (Athens, Ohio, 1975), p. 9.

emphasis on partisan politics, interest groups, and public opinion. This has remained a relatively unexplored link in the analysis of American politics and national security policy, despite the size and importance of the American national security establishment. It is the flip side of the national security coin that the preponderant influence of international relations on national security studies obscures, and that Americanists too rarely examine.[2]

Instead, the vast majority of work on defense has focused on two problems. Much of the defense literature, exemplified by the voluminous scholarship on nuclear deterrence, assesses the consequences of military policies, however they were formed, for national or international security. The other significant portion of security literature attempts to trace the causal influences on the formation of defense policy. Various theories, including systemic realism, rational actor models, psychological approaches, bureaucratic politics, economic influence, and public opinion contend in the search for explanatory power and precision.[3] In this body of work on the causes and international consequences of defense policy, elements of domestic politics, when considered at all, are usually treated as independent or intervening variables of secondary importance. The predominant tendency is to study policy makers' attempts to react and adjust to international events and circumstances and, on occasion, to include the ways in which those attempts are constrained by certain domestic factors. Even when some aspect of domestic politics figures importantly in the analysis, the emphasis is still on the implications for national security policy or international relations, not American politics.

The logic of the relationship between domestic political competition and national security policy is rooted in the fact that world politics imposes few unambiguous imperatives on statecraft. The per-

---

2. Works that include domestic political competition as an important part of their analysis include Alan Wolfe's *The Rise and Fall of the Soviet Threat* (Boston, 1984); Desmond Ball's *Politics and Force Levels* (Berkeley, 1980), in which, however, he emphasizes bureaucratic politics; Daniel Ellsberg's "The Quagmire Myth and the Stalemate Machine," in *Papers on the War* (New York, 1972), pp. 42–135; and Aliano, *American Defense Policy*. The lack of attention to domestic factors has been recognized by security studies specialists; see Joseph S. Nye and Sean M. Lynn-Jones, "International Security Studies," *International Security* 12 (Spring 1988): 5-27. There is a growing literature on, among other things, the role of public opinion in national security.

3. One highly regarded example of this kind of research is Barry R. Posen, *The Sources of Military Doctrine* (Ithaca, N.Y., 1984).

ception of external threats is in large measure subjective and open to multiple interpretations. The subjective perceptions of policy makers bestow meaning on international events and other security variables. These perceptions may be sincere or contrived, accurate or biased by imperfect information or psychological and ideological motivations. As Peter Gourevitch argues, although the international system may be a causal force in the formation of security policy, it is underdetermining: "The environment may exert strong pulls but short of actual occupation, some leeway in response to that environment remains. . . . The explanation of choice among possibilities therefore requires some examination of domestic politics."[4]

Because there is rarely an objective or compelling interpretation of international events and security requirements, the perceptions of policy makers provide latitude for political maneuvering. Domestic circumstances often influence a politician's beliefs on national security as much as any external threat, and external threats can be interpreted to harmonize with domestic needs, a connection aptly illustrated by the relationship of many members of Congress to local bases or military contractors. Perceptions and motivations are malleable, subject to revision according to internal political circumstances, rather than being governed solely or even primarily by a changing external environment. Consequently, the security implications of international constraints (such as the Soviet threat) are open to domestic political manipulation, within the confines of certain limitations (such as the cold war consensus, which was itself partly a product of domestic political competition). Moreover, any single perception or motivation can be addressed by more than one policy response. For example, the Soviet ICBM threat as perceived by the Reagan administration could have been met by several offensive options, but the administration chose strategic defense as its central effort.

Because perception of a security threat does not dictate any particular policy, and because perceptions can be altered to fit political needs, national security policy becomes a flexible political resource for politicians in their quest for power. By promulgating and marketing military policy initiatives or innovations, politicians attempt to enhance the relative power of their coalitions and allied institutions. Politicians can and do manipulate national security to mobilize con-

---

4. Peter Gourevitch, "The Second Image Reversed," *International Organization* 32 (Autumn 1978): 900.

stituencies and to undermine opposing politicians and coalitions, with the domestic consequences often far more profound than any international impact.

Implicit in this kind of study of defense and domestic politics is a particular approach to the analysis of American politics: through analysis of the role of political elites in shaping the issue cleavages and determining the scope of conflict in American politics, especially in their efforts to institutionalize political alignments and power.[5] This approach distinguishes itself from other theories of American politics by emphasizing the way politicians manipulate the political agenda and the scope of conflict, the role of political competition in recasting institutional power and coalitions, and the linkage between ideologies and political and economic interests. At its core is the premise that political factors shape political outcomes, that, in the words of Theda Skocpol, "State structures and party organizations have (to a very significant degree) independent histories. They are shaped and reshaped not simply in response to socioeconomic changes or dominant-class interests, nor as a direct side-effect of class struggles. Rather they are shaped and reshaped through the struggles of politicians among themselves, struggles that sometimes prompt politicians to mobilize social support or to act upon the society or economy in pursuit of political advantages in relation to other politicians."[6] National security policy provides a medium for such political struggle.

Thus, in contrast to some theories from international relations, politicians often have latitude for maneuvering within the constraints of perceived security requirements, and in contrast to many domestic perspectives, politicians can use defense policy to influence social and economic forces rather than simply being acted upon by society, industrial interests, and bureaucracies. This book presents a political analysis of the formation and consequences of na-

---

5. This approach draws on the theoretical and empirical insights from "realist" democratic theory and the study of political alignments and change in American politics. Some premises of this approach have theoretical roots in such works as E. E. Schattschneider, *The Semisovereign People* (Hinsdale, Ill., 1975), and Joseph Schumpeter, *Capitalism, Socialism, and Democracy* (New York, 1942). A more recent application of this approach is Benjamin Ginsberg and Martin Shefter's *Politics by Other Means* (New York, 1990); it has also been applied in a growing literature on the origins of political issues, for example, Edward G. Carmines and James A. Stimson's *Issue Evolution: Race and the Transformation of American Politics* (Princeton, N.J., 1989).

6. Theda Skocpol, "Political Response to Capitalist Crisis," *Politics and Society* 10 (1980): 200.

tional security policy, rather than assuming direct and decisive causal effects of social, economic, and international forces on defense politics and policy. Political interests change, and political elites can shape that change, with consequences for the institutions and coalitions they represent, especially during times of dealignment and realignment. To understand defense politics, it is necessary to understand the mechanisms and institutions through which both global events and social forces are organized into the debate over national security and the making of defense policy.

By examining the three defense policy agendas, the political coalitions behind them, and their interactions, this book investigates the relationship between defense policy and domestic politics during a period of significant political change. It seeks to illuminate the ways in which the fight over the size of the defense budget and the scope and purpose of the American arsenal profoundly influenced and continues to influence the course of government and politics in the United States. During the Reagan era, as in the past, the contest among domestic institutions and coalitions over the course of American military policy has played a major role in the process of political change and the formation of political coalitions. During the 1980s, debates over national security and innovations in defense policy often were shaped less by military imperatives, international events, or economic and bureaucratic interests than by the political strategies of elected officials and their opponents. In turn, these debates and policy initiatives greatly affected the dynamics of American politics.

In this book, although the three defense policy agendas and the resulting policy initiatives are treated in separate chapters, the organization is nonetheless largely chronological. The intent is to provide some of the focus of case studies while retaining much of the quality of a historical narrative. Chapter 1 gives a historical overview of the relationship between military policy and American politics from the beginning of the cold war era to the elections of 1980. Chapters 2, 3, and 4 describe the origins, political organization, and policy agendas behind the Reagan buildup, the peace movement, and the military reform movement. These chapters explore the connections between each defense policy movement and the political forces they represented in the larger domestic struggle, specifically: the use of the rhetoric and resources of national security to build and maintain Reagan's conservative coalition and undermine Democratic politics; the importance of the nuclear weapons peace movement in the mobilization of liberal opposition to the Reagan revolution; and the adoption and promotion of military reform, particularly by members of Congress, in response to the clash between the peace movement and the Reagan administration. The emphasis is on how the defense policy agendas were organized into national politics. Chapters 5 through 7 probe the political competition among these institutions and coalitions by examining their major policy initiatives—the nuclear weapons freeze proposal for the peace movement and the Democratic party, the Strategic Defense Initiative for the Reagan administration, and the attempts by the military reform lobby in Congress to change the Pentagon's procurement practices—and the impact of those initiatives on the defense debate and domestic politics. Chapter 8 summarizes the results of the conflicts at the end of Reagan's administration, the changes prompted by the revolutions of 1989, and the legacy of the Reagan era as the new decade begins.

1

Continuity and Change in Cold War Politics

The significance of national security politics in the Reagan era can be understood only in the context of changes over the past few decades in the institutions and coalitions of American politics generally and defense politics specifically. The transition in the politics of national security from the early cold war to the 1980s is part of a much broader evolution in American politics: the gradual dissolution of New Deal coalitions accompanied by the striving of parties and politicians to build new political alignments. Defense platforms and programs played an important part in the political competition that brought the United States from the zenith of New Deal alignments to the Reagan revolution.

Cold War Politics

All too quickly after World War II, the cold war began, and its impact on New Deal partisan competition was profound. Despite their general agreement on principles of defense and foreign policy, Democrats and Republicans found political disagreements about the conduct of the cold war a source of partisan struggle. From one perspective, the most telling characteristic of cold war national security politics, at least until the end of the Vietnam War, was the degree to which the political parties agreed on the ends, if not the precise means, of foreign and defense policy. Politicians and parties did not argue about the need to resist the Soviet Union and sociopolitical revolutions by maintaining a strong defense establishment and military alliances and, if need be, by the use of military force. Instead they often vied for the image of "most committed to military strength and resolve." Democrats and Republicans cooperated in building and maintaining the permanent military establishment. Yet until the 1970s the Democratic party was most closely associated with war, military buildups, and higher defense budgets.[1] An expanding postwar economy made possible the combination of welfare and warfare programs central to the maintenance of the Democratic coalition.

Although the cold war consensus limited partisan competition over national security policy largely to a debate among hawks, politicians found latitude within the consensus for politicizing issues of defense.[2] General agreement on ends or goals of foreign and defense policy did not preclude disagreement and competition over specifics of national security. Furthermore, consensus on the need to contain communism meant that the party in power would be held accountable for any real or perceived setbacks or defeats in containment. This blame game began with the onset of the cold war. Truman and the Democrats were held responsible for "losing China" in 1949. The Soviet testing of an atomic bomb that same year produced another setback, followed shortly by the Sino-Soviet alliance. Senator McCarthy began making his charges of communist infiltration of the State Department. Truman's popularity plummeted. Meanwhile, the president had authorized a study of national security strategy and requirements. The resulting National Security Council document—NSC-68—became the American declaration of cold war because of its unprecedented call for a massive and sustained military buildup in the absence of actual military conflict.[3] NSC-68 depicted a Soviet

---

1. Data on congressional partisan differences in support of military budgets and programs demonstrate that Democrats were more supportive of stronger defense in their voting from 1946 to 1960. Democrats in Congress attempted to increase the defense budget every year of Eisenhower's presidency except 1957. See Samuel P. Huntington, *The Common Defense* (New York, 1961), pp. 251–67.

2. For a similar outlook on this period see Gary W. Reichard, "The Domestic Politics of National Security," in *The National Security*, ed. Norman A. Graebner (New York, 1986), pp. 243–74. Reichard argues that the cold war consensus and bipartisanship were "born of political expediency" (p. 244).

3. Samuel F. Wells, "Sounding the Tocsin: NSC-68 and the Soviet Threat," *International Security* 4 (Fall 1979): 116–58. An early and detailed analysis of NSC-68 is Paul Y. Hammond's "NSC-68: Prologue to Rearmament," in *Strategy, Politics, and Defense Bud-*

political-military establishment bent on expansion and endowed with an improving military, especially nuclear, capability; the United States would have to more than double its financial commitment to national security if it were to avoid dire consequences in the near future. With some timely help from the North Koreans, the recommendations of NSC-68 became policy within a matter of months. The 25 June 1950 invasion of South Korea prompted a rapid and forceful military response from Truman, who was motivated in part by mounting domestic criticism.[4] The military budget for fiscal 1951 ballooned from a planned $13.5 billion to a final total of over $48 billion. Even after Korea, the defense budget never descended to the peacetime levels of 1947–50, after demobilization from World War II. The prolonged and difficult prosecution of the American "police action" in Korea allowed the Republicans and Ike to argue in 1952 that they would do a better job and bring the boys home.

Eisenhower maintained a commitment to a balanced budget, low taxes, and a cap on military spending. With some help from Khrushchev and de-Stalinization, he was able to quiet partisan conflict over cold war policy and limit defense expenditures during his first term. Indicative of Eisenhower's initial success at defense politics was Adlai Stevenson's major campaign issue on national security policy in 1956: the nuclear test ban. Stevenson argued that Eisenhower was not doing enough to pursue such a ban with the Soviet Union. The electoral failure of Stevenson's strategy had Democrats looking for another way to assault Republican policies, and not long into the second Eisenhower administration, the Republicans came under attack from a Democratic Congress and groups within the New Deal coalition for their reputed neglect of national security. Two events gave impetus to this partisan onslaught: the Gaither report and the launch of Sputnik.

Confronted with a Civil Defense Administration proposal for a blast and fallout shelter program at a cost of forty billion dollars, Eisenhower, hoping to derail the plan, appointed an eleven-member ad hoc committee of outsiders to evaluate civil defense requirements. The committee, headed by the chair of the Ford Foundation,

---

gets, ed. Warner R. Schilling, Paul Y. Hammond, and Glenn H. Snyder (New York, 1962), pp. 271–378. The full text of NSC-68 is reprinted in Steven L. Rearden, *The Evolution of American Strategic Doctrine* (Boulder, Colo., 1984), pp. 89–131.

4. Ernest R. May, in *Lessons of the Past* (New York, 1973), chap. 3, discusses the domestic pressures on Truman, along with other conditions that shaped his response.

H. Rowan Gaither, and encouraged by cold war Democrats among its members, expanded its inquiry into a thorough investigation of U.S. strategic and defensive requirements. As the Gaither committee deliberated, the Soviets conducted their first ICBM test in late summer 1957. In October, they launched Sputnik, the first successful satellite.

The resulting hysteria made America's putative decline in technological, educational, and military capabilities front page news for several months. The Gaither committee submitted its classified report soon after Sputnik. Leaks of its contents to the media prompted congressional Democrats to demand its public release. The report painted a grave and pessimistic picture of the balance between the strategic capabilities of the superpowers and called for substantial increases in spending on nuclear programs.[5] In particular, it argued that U.S. strategic bomber bases were becoming vulnerable to a preemptive strike. Consequently, it called for increased intermediate-range missile deployment in Europe and rapid acceleration of development and deployment of land-based and submarine-launched missiles. (This early version of the window-of-vulnerability argument would resurface in the 1970s, this time with land-based missiles as the endangered leg of the strategic triad and the Republicans as its most vocal exponents.) Borrowing some authors of the Gaither report, the Rockefeller Brothers Fund published a similar and equally pessimistic study in January 1958.[6] Several members of the Gaither committee went on to become advisors to John F. Kennedy's presidential campaign and then defense policy makers in his administration.

Democrats used their control over Congress to launch a series of attacks on the Eisenhower administration's restrained military budgets and programs.[7] Beginning in November 1957 with hearings by Lyndon Johnson's Senate Armed Services Preparedness subcommittee, congressional inquiries became Democratic platforms for demanding substantial increases in military spending, especially for

---

5. The report, entitled *Deterrence and Survival in the Nuclear Age*, was declassified in 1973 and printed for the use of the Joint Committee on Defense Production of the U.S. Congress (Washington, D.C., 1976). On the origins and consequences of the Gaither study see Morton Halperin, "The Gaither Committee and the Policy Process," *World Politics* 13 (April 1961): 360–84.

6. Rockefeller Brothers Fund, *International Security* (Garden City, N.Y., 1958). Henry Kissinger was a principal figure behind the study.

7. For a detailed history of the politics of the debate in this period see Richard Aliano, *American Defense Policy from Eisenhower to Kennedy* (Athens, Ohio, 1975).

nuclear programs. Traditional Democratic constituencies, including educational leaders, labor unions, and heavy industry, joined in the call for more defense dollars. Unable to ignore the pressures of the cold war consensus, a beleaguered Republican administration tried to deflect this assault by emphasizing, in part, military reform and reorganization. Eisenhower pushed what was clearly a subsidiary conclusion in the Gaither report: the need to strengthen unified decision making in military command and planning.[8] His proposed legislation eventually became the Department of Defense Reorganization Act of 1958. He did not implement the buildup called for by Democrats. The uproar over national security, and a faltering economy, contributed to the Democratic landslide in the congressional elections of 1958.

The following year the Democrats fielded in the campaign for the presidency a host of hawkish senators including Johnson, Kennedy, and Stuart S. Symington. At its convention the party united behind strident cold war rhetoric. The 1960 Democratic platform begins with and stresses the rebuilding of military might and the rollback of communism. Republican leadership had allowed U.S. military capabilities to fall behind to the point where "our military position today is measured in terms of gaps—missile gap, space gap, limited war gap." There was no debate over this part of the platform. The controversy was over a minority amendment, sponsored by several southern states, to strike the platform's language in favor of civil rights. In the debate, southerners, including Sam Ervin, made explicit reference to the South's importance to the Democratic electoral coalition: the party could not win without the South. In contrast to this divisive debate, national security provided the umbrella under which to unite the potentially fragile Democratic coalition.[9] Kennedy, the Democratic nominee, proceeded to campaign largely on the issues of the communist threat and national security.[10] Kennedy charged the Republicans with weakness and neglect while the Soviets were achieving military superiority. He made frequent refer-

8. In the Gaither report, changes in military organization are discussed under "Related Concerns." Eisenhower's call for reform just handed Democratic congressional leaders another vehicle for further inquiries into the state of national defense.

9. *Official Report of the Proceedings of the Democratic National Convention and Committee, 1960*, JFK Memorial Edition (Washington, D.C., 1964).

10. Kennedy used the communist threat occasionally to parry the issue of his Catholicism by citing national security as a more critical issue that united us all as Americans; see, for example, Theodore H. White, *The Making of the President, 1960* (New York, 1961), p. 468.

ences to the supposed missile gap and invoked the image of "the most critical period in our nation's history since the bleak winter at Valley Forge."[11] Kennedy edged out Eisenhower's vice-president, Richard Nixon, who had tried to distance himself from the administration's moderate policies. The Kennedy presidency ushered in another decade of Democratic cold war policies. Kennedy immediately added 15 percent to Eisenhower's final military budget. Emphasizing both nuclear deterrence and a conventional capability to intervene, President Kennedy implemented what would be the largest peacetime increases in defense spending until the Reagan era. The similarities between the military policies of Kennedy and Reagan go beyond increased spending. Both presidents had similar keystone military doctrines for conventional and nuclear armed forces to support the expanded budgets. Likewise, Kennedy's missile gap and calls for nuclear superiority were echoed by Reagan's window of vulnerability and plans for a nuclear "margin of safety." The missile gap of 1960 proved to be a combination of error and distortion. When Kennedy took the oath of office, the U.S. ICBM program had a considerable lead over its Soviet counterpart; whether he knew this during the campaign is uncertain.[12]

During these years a small but growing peace movement, which had rallied around the issues of the hydrogen bomb and atmospheric testing, sought entrance into a debate that barely recognized its premises and concerns. The peace movement, experiencing a renaissance after quiescent years in the late 1940s and early 1950s, expanded from growth in traditional pacifist organizations and the addition of professional, middle-class adherents, especially among scientists. Although its opposition to nuclear testing had some impact on the arms control policies of the Kennedy administration, the movement made no significant inroads into partisan and national politics. Kennedy's New Frontier came at the apex of American economic and military might. Prosperous times, the cold war consensus, and its own limited constituency kept the peace movement at the periphery of the national security debate.[13] Instead, Democratic cold war policies would leave as legacies the Vietnam War,

---

11. Quoted in Aliano, *American Defense Policy*, p. 231. Aliano provides many other quotations from the campaign, pp. 230–37. See also Desmond Ball, *Politics and Force Levels* (Berkeley, 1980), pp. 15–22.
12. Desmond Ball, *Politics and Force Levels*, pp. 21–25.
13. For a history of the peace movement in the late 1950s and early 1960s see Lawrence S. Wittner, *Rebels against War* (New York, 1969), chaps. 9 and 10.

which eroded the cold war consensus and precipitated the conversion of the Democrats into the more dovish of the two parties, and the nuclear missile arms race, which ultimately became the rallying point for a new peace movement twenty years later.

The Politics of Dealignment

Important events and developments during the two decades after Kennedy's election transformed American politics, setting the stage for the conflict described in this book. Partisan alignments changed measurably in the 1960s and 1970s as political constellations from the New Deal years began to shift.[14] Latent and suppressed tensions in the Democratic coalition emerged as certain issues developed. For example, civil rights solidified black allegiance to the Democratic cause as it precipitated white flight from the party, particularly by southerners. Gradually mounting opposition to the war in Vietnam forged stronger links among white liberals, students, and blacks, and their new coalition became a powerful force in the Democratic party. This alliance of liberal-left groups that emerged from the civil rights and anti-Vietnam movements, the so-called New Politics movement, found powerful expression in the open-government, environmental, women's, and consumer movements in the 1970s and pushed the Democrats toward more liberal positions on many issues, including intervention, military policy, and defense spending.[15] Partly as a consequence, the Democrats continued to lose their grip on their New Deal constituencies, including blue-collar and union laborers, urban ethnics, white southerners, and Catholics.[16]

As the Democratic party grew more dovish, the Republican party was developing strong pro–cold war constituencies. This coalition linked traditional sectors of GOP strength, including middle- and upper-class white suburbanites, to new sources drawn from disaffected ranks of largely Democratic constituencies. The support of

---

14. For one of several good analyses of the New Deal party system and its decline see Everett Carll Ladd, "The Shifting Party Coalitions," in *Party Coalitions in the 1980s*, ed. Seymour Martin Lipset (San Francisco, 1981), pp. 127–49.
15. The New Politics is discussed in greater detail in Chapter 3.
16. For quantitative evidence of group support over time for the Democratic party see Robert Axelrod, "Where the Votes Come From," *American Political Science Review* 66 (1972): 11–20; Axelrod, "Presidential Election Coalitions in 1984," ibid. 80 (1986): 281–84; John Petrocik, *Party Coalitions* (Chicago, 1981); Harold W. Stanley, William T. Bianco, and Richard G. Niemi, "Partisanship and Group Support over Time," *American Political Science Review* 80 (1986): 969–76.

these groups for cold war policies correlated positively with social conservatism; for many of those disgruntled by the Democratic party's adoption of the civil rights and women's movements also resented the loss in Vietnam and the apparent paralysis of American military might. These social forces and interests that supported Nixon and George Wallace in the late sixties and early seventies would be part of the broad coalition Ronald Reagan mobilized in 1980.

Even as the Vietnam War precipitated widespread social turmoil and intense partisan conflict, Richard Nixon's détente policies diminished other partisan differences over the cold war. The successful negotiation of the first Strategic Arms Limitation Treaty (SALT I) implied that arms control could be the way out of the nuclear dilemma. The opening of relations with China provided hope for a peaceful future as well as immediate political payoffs for the president. Vietnam eroded confidence in the ability of conventional military force to serve American interests and thus undercut public support for further interventions. Moreover, the relative decline, in the late 1960s and 1970s, of American economic performance combined with the oil price shocks to make the simultaneous expansion of both the welfare and warfare states more difficult. The winding down of American involvement in Vietnam produced decreasing defense budgets in harmony with public opinion, which wanted military cuts in favor of other domestic programs, including environmental protection, which was an emerging concern. Consequently, the cold war consensus that had regulated the competition over defense policy was partially dispelled.

Some of the other patterns of cold war politics were disrupted as well. As the party coalitions mixed and shifted, no dominant coalition emerged to replace the New Deal constellations, even in the wake of Vietnam and Watergate. In fact, Watergate probably derailed increasing Republican momentum. As a result, this period of change in party coalitions, including changes in defense politics, was part of an extended period of political dealignment. Dealignment, in turn, dovetailed with several other significant developments in American politics in this era with an important impact on defense politics in the 1980s. The decline, since the 1950s, of parties as an organizational influence in American politics was an important component of dealignment;[17] in fact, the decay of party power and

---

17. Walter Dean Burnham is foremost among analysts of the degeneration of the American party system; see his *Current Crisis in American Politics* (New York, 1982) for

partisan identification was among its first indicators. Dealignment was also manifest in political changes in the presidency and Congress.

During this era, presidential candidates developed their own popular coalitions, independent of party ties.[18] This made the presidency responsive to the social forces that have elected the Republican candidate every time but once since 1968, while the Democrats maintained majorities in both houses of Congress, except for six years in the Senate. Likewise, members of Congress acquired the resources and skills to run their own campaigns and minimize their dependence on party or president. Another part of dealignment, as a result, has been a consistent political separation between Congress and the president, a norm of divided government.[19]

The separation of president and Congress, exemplified by the Nixon era, exacerbated the institutional conflicts between the executive and legislature, and nowhere more than in defense and foreign policy. The power of the sometimes-imperial presidency stems in part from the vastly increased importance of national security and foreign policy in the postwar era.[20] The role of commander-in-chief in the nuclear era has been a potent source of presidential prerogative. Despite the loss of initiative to the executive branch, the Democratic Congress of the late 1960s and 1970s challenged presidential dominance, especially after Vietnam and Watergate. Perhaps the clearest manifestation of congressional resurgence and its challenge to presidential power has been a steady increase in congressional involvement in foreign and defense policy.[21] Although the political fragmentation of Congress circumscribes its ability to make coherent contributions to military policy, congressional knowledge and activism on national security issues has increased substantially. Defense became another area of issue specialization through which many

---

a compilation of his work on the subject. See also William J. Crotty and Gary C. Jacobson, *American Parties in Decline* (Boston, 1980); Martin P. Wattenberg, *The Decline of American Political Parties, 1952–1980* (Cambridge, Mass. 1984). A more optimistic view is provided by Xandra Kayden and Eddie Mahe, *The Party Goes On* (New York, 1985).

18. Theodore J. Lowi, *The Personal President* (Ithaca, N.Y., 1985).

19. Morris Fiorina, "The Presidency and Congress: An Electoral Connection?" in *The Presidency and the Political System*, ed. Michael Nelson, 2d ed. (Washington, D.C., 1988), pp. 411–34.

20. Arthur M. Schlesinger, Jr., *The Imperial Presidency* (Boston, 1973).

21. On congressional resurgence in the 1970s see James Sundquist, *The Decline and Resurgence of Congress* (Washington, D.C., 1981); James M. Lindsay, "Congress and Defense Policy," *Armed Forces and Society* 13 (Spring 1987): 371–400.

members could advance their legislative careers. Much of the defense policy debate in the 1980s was shaped by the new congressional politics and the separation between the two branches.

Finally, public interest groups emerged as increasingly powerful loci of political influence, especially with the declining influence of parties. Their rise owes much to the New Politics movement, which specialized in using these institutions to work its ends through Congress and the courts during the 1970s. The proliferation of public interest groups included an explosion of organizations, both liberal and conservative, devoted to matters of defense and foreign policy, and these groups reached out and mobilized a wider public. The nuclear freeze movement, as we shall see, was a powerful expression of this process of interest group mobilization.

All these changes and developments—including contradictions between welfare and warfare policies, weak parties and a trend toward divided government, interest group activism on foreign and defense policy, and the media's subsequent greater coverage of foreign and security affairs—further socialized the scope of conflict over defense policy and increased the integration of defense and domestic politics, thereby shaping the social and institutional framework for the conflict over defense policy in the 1980s.

Indeed, these developments manifested themselves quite forcefully in the rise and fall of the Carter presidency. In the fragmented and changing political environment of the middle 1970s, the election of Jimmy Carter appeared to consolidate Democratic power in the wake of both Watergate and Vietnam. The new president was elected by a liberal Democratic party, with a large base of voters and activists interested in revising traditional Democratic cold war policies. Carter pledged to reach arms control agreements, cut defense spending, and emphasize human rights in foreign affairs. One of the signal events early in his term was the cancellation of the B-1 bomber, which Carter announced on 30 June 1977. The cancellation, though it was the product of an alliance between the president, a coalition of liberal interest groups, and liberal Democrats in Congress,[22] nevertheless produced a bitter struggle in Congress. The president had to expend a great deal of his influence to line up members of his own party and ensure congressional sanction of the decision. Termination of the B-1 proved to be the high tide of Car-

---

22. On the B-1 cancellation see Nick Kotz, *Wild Blue Yonder: Money, Politics, and the B-1 Bomber* (New York, 1988), chaps. 12–13.

ter's new approach to military policy. Even before the cancellation, conservatives from both parties were organizing a counterattack against the politics of détente and the Vietnam syndrome.

Reconstituting Republican Power

In the middle years of the 1970s, conservative political elites set in motion a process that culminated in the election of Ronald Reagan and the emergence of the Republican party as the champion of cold war militarism. Whereas in 1960 a Democratic Congress and its presidential hopefuls had translated a Republican president's putative laxity on national security into political power, in 1980 Republicans would recapture the presidency and the Senate, in part through a similar attack on the Democratic chief executive. In 1960 the Democrats sought to rally their political constituencies around cold war policies that had been their forte since World War II. Two decades later, a movement of Republican forces would use the same issues and policies in an attempt to construct a political coalition and replace the decimated Democratic ranks as the dominant force in American politics.

It did not take long for conservative defense intellectuals and politicians to react to the damage done by Vietnam, détente, and Watergate to bipartisan foreign policies and to the power of the president—the commander in chief of the national security apparatus. Before Carter assumed that position, a coalition of cold war liberals, traditional Republican conservatives, and New Right Republicans began to construct an ideological and programmatic reconstitution of the cold war consensus. Conservative analysts and politicians revived the external threat essential to such a consensus. According to them, Vietnam and détente had obscured the public's perception of the Soviets as the clear and present danger. To make the threat credible they created and publicized revisions of the intentions, spending levels, and capabilities of Soviet military power. And because Vietnam and the recent attacks on the CIA's operations abroad had discredited the application of conventional and covert force, this cold war elite made the Soviet nuclear threat the focus of their campaign. These efforts established the framework of the national security debate during the 1980 elections and molded the policies of the Reagan presidency.

Leading the way for neoconservative activism was a newly formed alliance of defense intellectuals, businessmen, and politicians united as the Committee on the Present Danger (CPD). Organized in early 1976 by Eugene Rostow and Paul Nitze, the CPD had at its core many cold war Democrats,[23] and many of these nominal Democrats would soon throw their weighty influence against the Carter presidency and make alliances with Republican New Right forces. In contrast to the elite and intellectual organizations of the neoconservatives, the New Right was guided by direct-mail fund raisers, who began to flourish in the mid-1970s, with the founding of such groups as the National Conservative Political Action Committee (1975), the Conservative Caucus (1975), and the Committee for the Survival of a Free Congress (1974). Led by men like Richard Viguerie, Paul Weyrich, and Howard Phillips, these groups organized grass-roots conservative networks and accumulated vast mailing lists. They also were instrumental in persuading the electronic Baptist ministries to get involved in politics. This new cold war coalition of the neoconservative elites and the populist organizers would be central to the translation of the elite cold war consensus into a popular program for mass consumption.[24] As we shall see, with the campaign against the Panama Canal treaty and the SALT II treaty, issues of military policy and foreign policy, rather than social issues or domestic spending, became the initial common ground for the alliance that would be the backbone of the Reagan campaign.

Just after its inception the CPD and its allies were provided a powerful way to reassert the Soviet threat. Appointed director of the CIA at the same time the committee was organizing, George Bush, a man with political ambitions in an election year, made the reassessment of the National Intelligence Estimates (NIE) a top priority. The NIE serve as a basis for analysis of national security threats and defense program planning. President Ford agreed with Bush's idea of introducing a panel of explicitly hawkish or "pessimistic" "out-

---

23. Other executive committee members included Lane Kirkland, president of the AFL-CIO; David Packard, of Hewlett-Packard and former deputy secretary of defense; Dean Rusk, and Elmo R. Zumwalt, admiral, USN, ret. Among the well over one hundred board members were William Colby, John Connally, Nathan Glazer, Huntington Harris (a Brookings Institution trustee), Jeane Kirkpatrick, Seymour Martin Lipset, Norman Podhoretz, Matthew Ridgeway, Bayard Rustin, and Edward Teller.

24. Richard A. Viguerie, in his *New Right* (Falls Church, Va., 1981), details the rise of the New Right and its affiliations; Jerry W. Sanders also discusses the alliance and its impact on national security policy in his *Peddlers of Crisis* (Boston, 1983), chap. 6. For more general analysis of the growth of conservative power see Alan Crawford, *Thunder on the Right* (New York, 1980).

siders" to perform a "competitive analysis" of the Soviet threat along side the usual CIA team. Such an action was unprecedented. Team B, as it came to be known, was led by CPD member Richard Pipes and included founding members Paul Nitze and William Van Cleave.[25] Leaks to the press of Team B's classified findings in December 1976 revealed that they found standard CIA procedures to grossly underestimate the gravity of Soviet intentions, the scope of the Soviet military effort, and their military capabilities.[26] Team B argued, first, that the Soviets sought military superiority and were willing to use that advantage (as evinced by a supposedly massive civil defense program); second, that they had spent much more on military programs in the 1970s than we had at first estimated; and third, that their nuclear forces were becoming capable of a first strike against U.S. land-based missiles.

The parallels between the work of Team B and the findings and impact of NSC-68 in 1950 and the Gaither committee report seven years later are striking. All three presidentially sanctioned studies produced dramatic reassessments of the Soviet threat with important political consequences. A key player on Team B and the Gaither committee, Paul Nitze, was the principal author of NSC-68. He had just replaced George Kennan as director of the Policy Planning Staff at the State Department when Secretary Acheson initiated that study. Although the translation of Team B's analysis into policy would take four years (instead of four months for NSC-68), it provided immediate ammunition for political attacks on the supporters of détente and a declining defense budget. Then again, implementation of many of Team B's recommendations would not require a war, either; as with the Gaither study, a presidential election would suffice. However dubious some of the findings were, the revised NIE reflected Team B's work, received much publicity, and awaited examination by the incoming Carter administration. In this way, just as NSC-68 and the Gaither report did in previous decades, Team B

---

25. Other members were General Daniel Graham, former head of the Defense Intelligence Agency; Thomas W. Wolfe, of the Rand Corporation; Paul D. Wolfowitz from the Arms Control and Disarmament Agency; and John Voght, a retired air force general.

26. David Binder, "New CIA Estimate Finds Soviets Seek Superiority in Arms," *New York Times*, 26 December 1976, pp. 1 and 14, sec. A. For later accounts of the process and meaning of Team B's work see Bruce Cummings, "Chinatown," in *The Hidden Election*, ed. Thomas Ferguson and Joel Rogers (New York, 1981), pp. 196–231; and Robert Scheer, *With Enough Shovels* (New York, 1983), chap. 5.

helped set the terms of the national security debate from 1977 to 1980 and beyond.[27]

The conservative elite made nuclear weapons and strategy the centerpiece of the revived Soviet threat. Beyond the well-publicized governmental work of Team B, an intellectual campaign filled journals with articles that juxtaposed aggressive Soviet nuclear capabilities and strategy with American vulnerabilities and dereliction of strategic forces. Paul Nitze is often credited with being the father of the fundamental argument of this campaign: the window of vulnerability argument. He and others argued, first, that Soviet land-based missiles, ceteris paribus, would at a point in the near future be capable of destroying the U.S. ICBM forces, thereby giving the Russians leverage for international blackmail, if not victory in nuclear war.[28] Again, the logic and premonition of the argument are reminiscent of the "year of maximum danger" predicted by Nitze's NSC-68. Second, the Soviets not only had the capability, but also the political and military will and strategy, to carry out such an attack.[29] Finally, much was made of the programmatic and monetary neglect of strategic programs during the Vietnam years and the failure to amend our relative decline during the period of détente.[30]

All these arguments became grist for the political and intellectual mills of conservative political advocacy groups and think tanks that experienced a renaissance during these years, thanks, in no small part, to the political mobilization of American business and the re-

---

27. Arthur Macy Cox, "The CIA's Tragic Error," *New York Review of Books*, 6 November 1980, pp. 21–24. In addition to recounting the legacy of Team B, Cox focuses on the misinterpretation of the CIA's reestimation of Soviet military spending. The revision upward of Soviet expenditures from 6–8 percent to 11–13 percent of GNP should have suggested not that the Soviet military buildup had been greater than originally thought but that their military-industrial complex was less efficient than estimated. The Soviets must have spent more to achieve the same observed output. Nevertheless, the distortion of this reevaluation became the basis for Reagan's repeated references to Soviet military spending in the 1970s as the largest buildup in history. The pattern of official threat reassessments as precursors of each buildup has been discussed by others, including Alan Wolfe, *The Rise and Fall of the Soviet Threat* (Boston, 1984).

28. Paul Nitze, "Assuring Strategic Stability in an Era of Detente," *Foreign Affairs* 54 (January 1976): 207–32; Nitze, "Deterring Our Deterrent," *Foreign Policy* (Winter 1976–77): 195–210. For a critique of the Nitze position see Jan M. Lodal, "Assuring Strategic Stability," *Foreign Affairs* 54 (April 1976): 462–81.

29. Richard Pipes, "Why the Soviet Union Thinks It Could Fight and Win a Nuclear War," *Commentary*, July 1977, pp. 21–34.

30. For a classic example see Albert Wohlstetter's piece, which concluded a debate in *Foreign Policy*, "Optimal Ways to Confuse Ourselves" (Fall 1975): 170–98.

sulting generous increases in corporate funding.[31] These institutions, including the American Enterprise Institute, the Hoover Institute, the Heritage Foundation, and the American Security Council, further publicized and popularized these themes, which would become the stock iterations of conservative Republican candidates in general and of Ronald Reagan in particular.

But before these arguments came to benefit Reagan in 1980, the conservative advocacy and political action groups used them to discredit the Carter administration's initial foreign and defense policies. Attacks on Carter's foreign and military policies began almost immediately, with a campaign to foil the president's nomination of liberal Paul Warnke to head the Arms Control and Disarmament Agency and the SALT delegation.[32] Next came the attempt to stop ratification of the Panama Canal treaty. This took place in 1978, which Viguerie called the "critical year" for the New Right mobilization.[33] New Right political action groups launched an expensive campaign to turn public opinion against the treaty. During the Panama Canal debate, the American Conservative Union, by itself, spent $1.4 million on lobbying, sent 2.4 million pieces of mail, and bought thirty minutes of TV time on over three hundred stations.[34] The fight fostered alliances between the New Right and some members of Congress, such as Paul Laxalt. It also set the stage for the election of some conservative Republicans like Gordon Humphrey and Roger Jepsen that November. At each step, the links grew stronger between the elite neoconservative network and the New Right populist organizers. The neoconservative elite benefited from the popular pressure applied on behalf of their causes. The Vigueries and Weyrichs gained credible arguments and respectability for their national security platforms. This alliance all but destroyed Warnke's credibility and effectiveness, and it nearly derailed the Panama Canal treaty.

---

31. Michael R. Gordon, "Right-of-Center Defense Groups: The Pendulum Has Swung Their Way," *National Journal*, 24 January 1981, pp. 128–32; Ann Crittendon, "The Economic Wind's Blowing toward the Right, for Now," *New York Times*, 16 July 1978, p. A19. On the general resurgence of corporate political clout in the late 1970s see David Vogel, "The Power of Business in America," *British Journal of Political Science* 13 (January 1983): 19–44; Thomas Byrne Edsall, *The New Politics of Inequality* (New York, 1984), chap. 3.

32. For fuller treatments of this and the other conservative attacks see Sanders, *Peddlers of Crisis*.

33. Viguerie, *New Right*, p. 65.

34. *Congressional Quarterly Weekly Report*, 22 April 1978, p. 952. Viguerie claimed that the canal debate added four hundred thousand names to his list, making the canal a no-lose issue for the New Right.

This escalating conservative effort culminated in the massive campaign to prevent ratification of SALT II. Where the alliance had fallen just short in its earlier efforts, this time it succeeded. The key group behind the attack was the Coalition for Peace through Strength, organized in 1978 by the American Security Council (ASC).[35] The coalition exemplified the growth of the conservative movement in scope and depth. The group announced itself in early August 1978, the day the House passed a defense bill authorizing an aircraft carrier opposed by Carter, thus marking the beginning of congressional increases in the administration's military programs and budgets.[36] Although attention focused on the nearly two hundred members of Congress from both parties who joined, the coalition also brought together industrialists, other conservative elites, and New Right activists. The coalition coordinated the campaign, which featured lobbying, direct mail, and heavy use of the mass media. The administration, and the State Department in particular, responded with their own public relations campaign.[37] The Carter administration failed to build on initial support for the treaty, while the conservative coalition mobilized opposition (Table 1). When the USSR invaded Afghanistan in late December 1979, Carter, knowing he lacked the necessary votes, withdrew the treaty from

Table 1. Informed opinion on SALT II (in percentage of survey participants)

	March 1979	June 1979	September 1979	March 1980
Informed on the issue	45	64	61	61
Proratification	27	34	24	26
Antiratification	9	19	26	26
No opinion	9	11	11	9

Source: Gallup Opinion Index 176 (March 1980).

35. The ASC began in 1955 as a McCarthyite, anti–internal subversion group; in this conservative mobilization, it evolved into a prodefense lobby that grew rapidly in membership and finances in the middle and late 1970s, enabling it to help bankroll the founding of the Center for International and Strategic Studies at Georgetown. Crawford, *Thunder on the Right*, p. 34.
36. Richard Burt, "Pro-Arms Coalition Formed in Congress," *New York Times*, 9 August 1978, pp. A1, D19; *Congressional Quarterly Weekly Report*, 12 August 1978, p. 2099; Sanders, *Peddlers of Crisis*, pp. 223–28.
37. Alan Berlow, "Emotional and High Priced Lobby Campaign Underway on Arms Limitation Treaty," *Congressional Quarterly Weekly Report*, 23 June 1979, pp. 1210–19.

Table 2. Public opinion on defense spending, 1974–81 (in percentage of survey participants)

	1974	1976	1977	1979	1981
Too little	12	22	27	34	51
Too much	44	36	23	21	15
About right	32	32	40	33	22
No opinion	12	10	10	12	12

Source: Gallup Opinion Index 175 (February 1980) and 186 (March 1981).

the process of advice and probable dissent. The Soviet invasion of Afghanistan was not the cause of SALT II's demise; it merely provided Carter with an excuse to withdraw the stillborn treaty from senate consideration.

Fostered in part by the conservative campaign, the tide of public opinion had turned by late 1979, not only against SALT II, but also decidedly in favor of increased militarism and military spending. In less than four years, from 1976 to 1980, public opinion on defense spending was turned around from a strong plurality that believed the United States was spending too much on defense to a near majority in favor of greater defense spending (Table 2). By late 1979, Carter had committed himself to 5 percent annual increases in military spending.

With elections drawing nigh, the Carter administration initiated a series of eleventh-hour cold war policies. The invasion of Afghanistan ostensibly prompted several get-tough decisions. Besides withdrawing SALT II, Carter resumed draft registration for young men. He added a grain embargo and a boycott of the Moscow summer Olympics. Carter's increases in defense spending only induced further additions from the Democratic Congress: in 1980 it increased a president's military appropriation request for the first time in thirteen years.[38] Republicans harped on Democratic laxity with regard to the Soviet menace and other putative examples of America's decimated power, such as the Iranian hostage crisis. Election-year retrenchment by the Democrats pleased neither hawk nor dove. Even Carter's flurry of nuclear posturing, including the public announcement of a supposedly new and more muscular nuclear deterrence

---

38. Congressional Quarterly, *U.S. Defense Policy* (Washington, D.C., 1983), p. 25.

strategy, failed to deflect charges of weakness. By this time, however, the neoconservative elite and the New Right populists had sealed their alliance. Perhaps most symbolic of this union of cold war Democrats and the populist wing of the Republican party was the induction of Ronald Reagan into the executive council of the CPD in the fall of 1979.[39]

Reagan, the Republicans, and the Elections of 1980

Ronald Reagan led the GOP presidential hopefuls in a relentless campaign chorus of contempt for Carter's foreign and defense policies, a contempt barely superseded by the rhetoric directed against the Soviet Union. A decade of détente, they said, had let America's defenses decay while Soviet military power grew. Whereas the Democratic party gathered for its convention divided between moderates and cold warriors, the Republicans had little trouble speaking in one voice. The GOP emerged united behind a massive military buildup. The alarmist and strident tone of the platform was virtually unprecedented: "Overseas conditions already perilous, deteriorate. The Soviet Union for the first time is acquiring the means to obliterate or cripple our land-based missile system and blackmail us into submission. Marxist tyrannies spread more rapidly through the third world and Latin America. . . . if we let the drift go on . . . the American experiment, so marvelously successful for 200 years, [could come] strangely, needlessly, tragically to a dismal end early in our third century."[40] The unusually long section titled "Peace and Freedom" begins with the warning that "at the start of the 1980s the United States faces the most serious challenge to its survival in the two centuries of its existence."[41] This "most serious challenge" was not evident to the incumbent party just four years earlier at its convention in 1976, when its national security plank was about one-quarter the size and devoid of alarmist rhetoric. Meeting the challenge in the 1980s required a military buildup that would "ultimately reach the position of superiority that the American people demand."[42] Already cultivated by the cold war coalition's campaign, a receptive

---

39. *National Journal*, 20 October 1979, p. 1751.
40. *National Party Platforms of 1980*, comp. Donald Bruce Johnson (Urbana, Ill., 1982), p. 176.
41. Ibid., p. 204.
42. Ibid., p. 206. In contrast, the 1980 Democratic defense plank is reminiscent of the 1976 GOP plank—calmly committed to a defense second to none, but also concerned about spending wisely.

American public grew more hawkish as the Iranian hostage crisis dragged on into the fall of the election year.

To maximize the electoral payoffs from the general rise in conservative and prodefense attitudes, Republican campaign strategists targeted "key demographic groups," including union members, blue-collar workers, Catholics, Hispanics, and middle-aged voters.[43] The Reagan pollsters determined that these groups embodied the values of traditional ethics, a respect for strong leadership, and an America-first vision.[44] Despite Carter's regional identity, the GOP also wooed white southerners, having identified them as among the most nationalistic and supportive of forceful foreign and military policy. The strategy worked. Significant numbers of all these groups shifted from support of Carter in 1976 to Reagan in 1980 (Table 3).

The question of how much impact issues of national security had on the electoral fortunes of Reagan and Republicans in 1980 cannot be assessed with great confidence; public opinion data offer few insights.[45] The Iranian hostage crisis, which certainly did evoke public concerns about American power and foreign policy, tended to swamp broader insights into public concerns about defense and national security. The economy was the other focal issue and may have had more to do with Reagan's victory than purely military issues.[46] Nevertheless, most analysts concluded that in 1980, Reagan gained a distinct electoral advantage from issues of foreign policy and national security.[47]

Election day produced the broadest gains for the Republican party

---

43. Richard Wirthlin, "The Republican Strategy and Its Electoral Consequences," in *Party Coalitions in the 1980s*, ed. Seymour Martin Lipset (San Francisco, 1981), p. 238. Wirthlin was candidate Reagan's pollster and a central campaign strategist.

44. Ibid., pp. 242–43.

45. Very few questions on the Center for Political Studies, CBS/*New York Times*, or any other poll even indirectly tapped this question. For example, the CBS/*New York Times* 1980 election day survey asked, "Which issues were most important in deciding how you voted today?" Respondents were allowed a first and second mention, but from these choices: balancing the budget, crisis in Iran, jobs and unemployment, reducing taxes, ERA and abortion, inflation and economy, needs of big cities, and U.S. prestige around the world. Only "crisis in Iran" and "U.S. prestige" are linked to the question of national security and military strength, but not directly. The analogous CPS question is open ended; so many responses and so few people per response make coherent interpretation impossible.

46. In October 1979, Gallup polls had only 6 percent of Americans listing international events and foreign policy as the most important problem facing the nation (versus 63 percent for inflation). By January 1980, shortly after the embassy takeover, foreign policy jumped to 44 percent (versus 39 percent for inflation). *Gallup Opinion Index* 175 (February 1980).

47. Herbert Asher, *Presidential Elections and American Politics*, 3d ed. (Homewood, Ill., 1984), p. 164.

Table 3. Changes in voting for president, 1976 and 1980 (in percentages)

	1976		1980		Shift to GOP[a]
	D	R	D	R	
Whites	48	52	40	60	8
White southerners	47	53	36	64	11
Catholics	55	45	44	56	11
Blue-collar workers	58	42	49.5	50.5	8.5
Union members	60	40	52	48	8
Hispanics	76	24	60	40	16
Those aged 45–59	48	52	42	58	6

Sources: CBS News/*New York Times* Surveys, reported in the *New York Times*, 9 November 1980, p. 28, and *Public Opinion* (December–January 1985): 34.

[a] These shifts to the GOP are well above the average for all demographic categories given in the survey.

since Eisenhower. Ronald Reagan had crushed Carter as well as John Anderson. Republicans controlled the Senate for the first time since the elections of 1954 ended a one-term majority. They also made substantial gains in the House and at the state level. Regardless of the exact sources of his majority, Ronald Reagan was elected with strong support for a military buildup, a more aggressive foreign policy, tax cuts, and reductions in domestic programs. The Republicans seemed as united as the Democratic party seemed befuddled and divided. Subsequent discussion of political realignment flooded journalistic and academic assessments of the election. While one election could only portend a fundamental alteration in political power, 1980 did commence a relatively dramatic transformation of domestic and military policies. We now must look at how military policy fit into the Reagan revolution as a whole and how the Reagan administration used military policy and programs in an attempt to consolidate power. The elections of 1980 initiated a massive military buildup and a decade-long battle over defense policy as a pivotal element in the larger struggle over the future of American politics.

2

Military Policy and the Reagan Revolution

The election of Ronald Reagan ushered in a new era of cold war tension and commenced the most significant changes in American politics since the New Deal. During the campaign of 1980, Reagan and the Republicans had only ideological appeals and promises with which to gain support—promises to, as they saw it, restore U.S. military might, cut unnecessary domestic programs and spending, and lower taxes. Once in office, Reagan and his allies began to implement that platform. Never before in peacetime had defense policy figured so prominently in the construction and maintenance of a presidential coalition and a public policy agenda. Defense policy served as a potent and flexible political resource for the Reagan regime. The rhetoric, programs, and budgets of national security had powerful political and economic effects. In particular, increased defense spending became the linchpin of the Reagan administration's political and fiscal squeeze on the domestic priorities and programs critical to the Democratic party. For this and other reasons, the military programs of the Reagan administrations had less effect on the prospects for war and peace among nations than they had on domestic politics and economics in America.

A New Cold War and the Defense-Spending Buildup

To borrow the terminology of John Lewis Gaddis, Reagan's national security policy was a return to symmetrical containment; that is, national security policy based on the primacy of military means and the general expansion of military forces across all areas.[1] Symmetry also entails a strong disposition to perceive Soviet influence wherever international conflict occurs and to confront Soviet influence directly with force wherever it is perceived. From this perspective, global politics is a zero-sum game, played along an East-West axis, with military forces eternally trump. Although all presidents and administrations adhered to the principle of peace through strength, not all presidents believed that principle necessitated an upward spiral of defense spending. For various reasons, Eisenhower, Nixon, and Carter (initially) attempted to limit military expenditures and emphasize other tools of international politics, such as diplomacy. Under Reagan, however, the durable maxim "si vis pacem, para bellum" would take on new meaning and reality.

While every administration had indulged sporadically in Kremlin bashing, the Reagan White House sustained an unprecedented rhetorical barrage, nowhere better exemplified than in President Reagan's famous judgment that the Soviet Union was the "evil empire" and the "focus of evil in the modern world."[2] And as the president argued in his first press conference, the Soviet government reserved "the right to commit any crime, to lie, to cheat."[3] To make matters worse, candidate and President Reagan believed that the United States had slipped into a position of military inferiority during the 1970s. As he stated more than once, "In military strength we are already second to one; namely, the Soviet Union."[4] In a March 1982 press conference the president reaffirmed that "on balance the Soviet Union does have a definite margin of superiority."[5] Reagan de-

---

1. In his *Strategies of Containment* (New York, 1982), Gaddis contrasts symmetrical containment with asymmetric containment, which relies on a more balanced variety of means, including economic, diplomatic, and psychological measures.
2. "Excerpts from President's Speech to National Association of Evangelicals," *New York Times*, 9 March 1983, p. A18.
3. Transcript of President Reagan's first news conference, *New York Times*, 30 January 1981, p. A10.
4. From a 17 March 1980 campaign speech, quoted in "A Buildup in U.S. Forces," *Washington Post*, 16 June 1980, p. A1.
5. Quoted in Strobe Talbott, *Deadly Gambits* (New York, 1985), p. 6. Weinberger and other administration officials cited Soviet superiority in congressional testimony in support of their strategic programs. For example, in July 1981, Edward Rowny,

scribed the years before his presidency as a time when the Soviet Union "engaged in the biggest military buildup in the history of man at the same time that we tried the policy of unilateral disarmament, of weakness, if you will. And now, we are putting up a defense of our own."[6]

This Soviet advantage was most evident in the putative land-based missile window of vulnerability that left the United States exposed to, if not an outright first strike, then nuclear blackmail—military coercion on the basis of obvious Soviet strategic superiority. No shots need be fired; victory and defeat could be determined as in a game of chess. As Reagan put it, if the window were open wide enough, the Soviets "could just take us with a phone call."[7] Deterrence was a matter of capabilities and, more important, the perception of those capabilities. For Reagan and his followers, there could be no predefined sufficiency of military might because deterrence is a dynamic process of balancing incentives and counterincentives.[8] A nation could not be merely strong; it had to deter even irrational temptations. This implied that mutual assured destruction (MAD) or any notion of sufficiency was insufficient and imprudent. Qualitatively and quantitatively superior forces had to be matched to a more exacting strategic doctrine. Preventing the variety of conceivable nuclear Pearl Harbors was a demanding process. This notion of deterrence, supported by the window-of-vulnerability argument, was Reagan's principal agitprop for his nuclear buildup.

Just as Pearl Harbor was a haunting and useful image for many cold war policy makers, even more was Munich. The lesson of the Munich agreement in 1938 was that appeasement only encourages aggression and that perceived military weakness will be exploited by enemies. Reagan invoked the image of Munich several times.[9]

---

soon to be the president's chief negotiator for strategic arms control, testified that official reports demonstrated that "somewhere between the last quarter of 1980 and the first half of 1981, the Soviets did surpass us in overall strategic superiority"; quoted in Phil Williams and Stephen Kirby, "The ABM and American Domestic Politics," in *Antiballistic Missile Defense in the 1980s*, ed. Ian Bellany and Coit Blacker (London, 1983), p. 57.

6. From the second 1984 presidential debate in Kansas City, *New York Times*, 22 October 1984, p. B4.

7. Robert Scheer, *With Enough Shovels* (New York, 1983), p. 66.

8. For example, see one of Defense Secretary Caspar Weinberger's myriad disquisitions on the requirements of deterrence and the logic of peace through strength: "U.S. Defense Strategy," *Foreign Affairs*, 64 (Spring 1986): 675–97.

9. For a critical discussion of the lessons of Munich and how those lessons have influenced U.S. policy makers including Reagan, see Robert J. Beck, "Munich's Lessons Reconsidered," *International Security* 14 (Fall 1989): 160–91.

Munich's corollary is that negotiations can only be pursued from a position of military strength or advantage. Munich implied that a buildup had to precede negotiations and that only the product of the buildup would extract concessions and compliance. No agreement with the Soviets should leave even the impression of a Soviet advantage. Only once a "margin of safety" was restored could negotiations proceed. As candidate Reagan put it, "Once we clearly demonstrate to the Soviet leadership that we are determined to compete, arms control negotiations will again have a chance."[10]

But a more fundamental problem remained: if the Soviets reserved the right to commit any crime, to cheat, to lie, then bilateral agreements were hardly worth pursuing. Here, the historical legacy of the Yalta conference of 1945 held sway. The prevailing American interpretation of Yalta is that Stalin reneged on agreements to allow free elections in Poland, in particular, and elsewhere in Eastern Europe. Yalta thus complicated or complemented the Munich syndrome by providing evidence that the Soviet Union probably could not be trusted regardless of American military might. Consequently, arms control agreements, even if reached from a position of strength, were tenuous and suspect. Implicitly, the lessons of Pearl Harbor and Yalta instilled the common fear among conservative policy makers that Soviet cheating under any arms control agreement could lead to Soviet "breakout" from the treaty once the Soviet Union had surreptitiously achieved qualitative and quantitative superiority.[11]

This revised doctrine of peace through strength and the astringent cold war rhetoric were matched by an extraordinary commitment of resources. The primary measure of military might was in dollars spent. Money was the principal measurement of the putative decade of neglect in the 1970s and of the American commitment to security.[12] Defense Secretary Caspar Weinberger put the choice bluntly: "This

10. Quoted in "A Buildup," *Washington Post*, 16 June 1980, p. A4. Or as Paul Nitze, soon to be Reagan's INF arms control negotiator, put it, "There could be serious arms control negotiations, but only after we have built up our forces"; quoted in Scheer, *With Enough Shovels*, p. 90.

11. Many members of the administration, such as Richard Perle and Richard Burt, were openly hostile to the arms control process. Reagan himself made frequent reference to a book by a personal friend, Laurence Beilenson, titled *The Treaty Trap*, which said that, historically, nations did not survive by relying on paper agreements.

12. Defense Secretary Weinberger described the decade of neglect "in which US defense efforts declined by 20 percent in the face of a 50 percent increase in Soviet military strength" and said the election of 1980 was an affirmation that the United States "agreed to pay the price for military strength to deter war." Weinberger, "U.S. Defense Strategy," pp. 696 and 675.

choice is ours: we can buy the forces required to secure freedom and peace for ourselves, our allies, and our descendants; or we can meanly conclude it is too great an effort, falter and thus yield to the forces of totalitarianism and tyranny."[13] The price tag for freedom and peace was staggering. The Reagan Pentagon envisioned a total expenditure of $2.7 trillion on defense from 1982 through 1989, with a projected military budget of nearly $450 billion for fiscal year 1989. Reagan and the Pentagon would end up requesting over $430 billion less than they projected, and Congress eventually cut nearly $150 billion from that. Nevertheless, Reagan sustained the largest buildup in peacetime history, which by many measures exceeded spending during the Korean and Vietnam wars.

The single largest leaps in defense spending came in the early months of the administration, with the additions to Carter's FY 1981 budget and the formulation of the FY 1982 budget. The strategic doctrine and weapons programs of the Reagan buildup demanded massive increases in defense spending without, however, specifying the precise dollar figures as rational planning should. Much of the 1981 and 1982 increase was therefore the product of bargaining among White House advisors, who proceeded to debate how much to add to Carter's 1981 budget of nearly $160 billion, which was already a substantial increase over 1980, and to his proposed 1982 budget, which added even more real growth. Defense Secretary Weinberger got agreement on adding a total of $32.6 billion, an enormous increase over and above the approximately $20 billion of real growth in the Carter budgets. The result was 12.5 percent real growth in military budget authority in 1981 and 12 percent in 1982. Although Weinberger claimed that the number was a product of the Pentagon review process, the figure, emerging in less than one month from Reagan's inauguration, preceded any such review and took the Pentagon by surprise. The services were compelled to find programs to justify the increase.[14] As for annual real increases beyond 1981 and 1982, Weinberger and David Stockman apparently haggled over the percentage and settled the matter within an hour

---

13. U.S. Department of Defense, *Report of the Secretary of Defense Caspar W. Weinberger to the Congress on the FY 1987 Budget, FY 1988 Authorization Request and FY 1987–1991 Defense Programs*, 5 February 1986, p. 5.

14. Nicholas Lemann, in "The Peacetime War," *Atlantic Monthly*, October 1984, p. 72, quotes several Pentagon and Office of Management and Budget employees about their having to scramble to find low-priority programs to soak up the increases; see also Richard Stubbing, *The Defense Game* (New York, 1986), pp. 374–75. Stubbing worked in the OMB on defense planning during this period.

on 30 January 1981, just ten days into the Reagan presidency. Weinberger wanted 9 percent per year; Stockman's plans were more toward 5 percent; and they settled on 7.[15]

In current dollars, Reagan's outlays of well over $1 trillion in his first five years nearly equaled the total spent by Nixon, Ford, and Carter over the course of their combined twelve years in office. Military budget authority doubled in just five years, from $143.9 billion in 1980 to $294.7 billion in 1985 (Table 4). In inflation-adjusted dollars this represents a real increase of 54.7 percent.[16] Outlays, or money actually spent per year over the same period, increased by 40 percent in real terms, reaching 48 percent by 1986. The outlays in FY 1986 exceeded in constant dollars those of any military budget since 1946, except for two peak years during the Korean (FY 1953) and Vietnam (FY 1968–69) wars. Military budget authority grew in real terms for five consecutive years, from 1981 to 1985 by an average of 9.3 percent. Outlays increased for six consecutive years, from 1981 to 1986 by an average of 6.8 percent. Never before in the postwar era had military budget authority or outlays increased for more than three years running. The 6.8 percent average annual increase in outlays greatly exceeds the average annual growth of just over 2 percent from 1964 to 1970, the core of the Vietnam War years. Had

Table 4. Military spending, fiscal years 1980–86 (in billions of current dollars)

	1980	1981	1982	1983	1984	1985	1986
Budget authority ($)[a]	143.9	180.0	216.5	245.0	265.2	294.7	289.1
Current $ increase		36.1	36.5	28.5	20.2	29.5	−5.6
% Real growth[b]		12.5	12.0	9.1	5.3	7.5	−4.2
Outlays ($)[c]	134.0	157.5	185.3	209.9	227.4	252.7	273.4
Current $ increase		23.5	27.8	24.6	17.5	25.3	20.7
% Real growth[b]		4.5	8.1	8.6	5.0	8.6	5.9

Source: Office of Management and Budget and Center on Budget and Policy Priorities, "The FY 1988 Defense Budget: Preliminary Analysis," 6 January 1987. All budget figures are for budget category (O50) National Defense.

[a]The total amount authorized for programs to be spent that year and in future years.

[b]Percentage growth from the previous year in inflation-adjusted dollars.

[c]The total amount actually spent during that year including budget authorizations from previous years and the current year.

15. David Stockman, *The Triumph of Politics* (New York, 1987), pp. 116–19.

16. All figures are taken or calculated from budgetary data provided by the Department of Defense and the Office of Management and Budget.

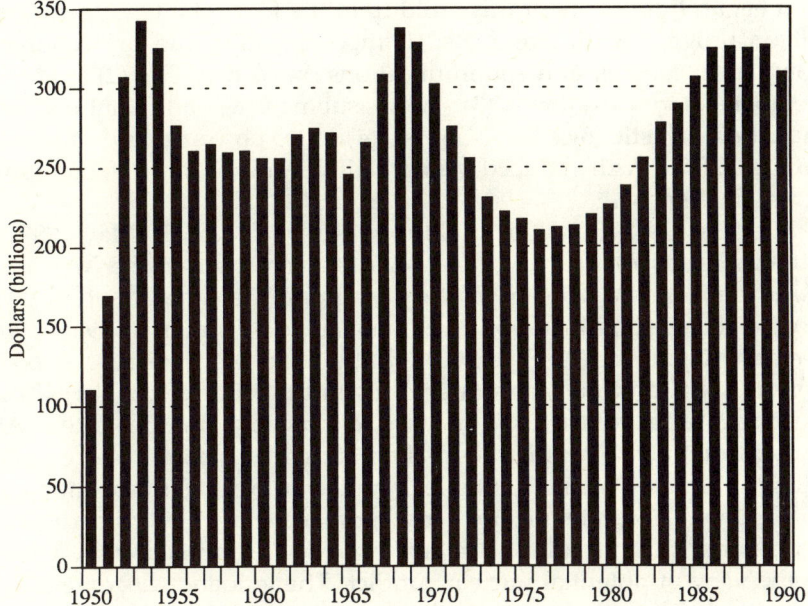

Figure 1. Military spending, 1950–1990 (in constant 1991 dollars)

Reagan controlled both houses of Congress, the buildup would have been more pronounced. Whereas Congress had inflated the final defense budgets of the Carter administration, it decreased Reagan's requests after FY 1982, his first complete budget year. From FY 1982 through FY 1985, Congress trimmed just over $69 billion, or an average of $17 billion for each of the four years.

This extraordinary surge in defense spending was devoted to the modernization and expansion of the gamut of military programs, conventional and nuclear, but first and foremost to the nuclear weapons rearmament program. As a presidential candidate, Reagan's charges of America's military woes and Soviet might focused on nuclear weapons, with repeated references to a nuclear window of vulnerability and Soviet superiority in strategic nuclear forces. He and others in the Republican cohort, such as George Bush, intimated that nuclear wars could be fought and won.[17] Behind this rhetoric was the clear intention to implement an enormous strategic nuclear weapons buildup. Upon assuming power, the administra-

---

17. See Scheer, *With Enough Shovels*.

tion began to implement this buildup in the form of a five-year, $180 billion program with five goals: (1) modernization and protection of command, control, and communications (known as C^3), (2) strategic bomber modernization, (3) new submarines and submarine-launched ballistic missiles (SLBMs), (4) more powerful and more accurate ICBMs with reduced vulnerability, and (5) improvements in strategic defense.[18]

The strategy behind this buildup was revealed on 30 May 1982, a little more than a year into Reagan's first term, when the *New York Times* published a front-page story outlining the contents of an otherwise-secret administration document titled "Fiscal Year 1984–1988 Defense Guidance."[19] Even though the document discusses both conventional and nuclear strategy, given the tenor of the times, the headlines and controversy focused exclusively on the latter. Although previous administrations had subscribed to aggressive operational nuclear doctrines, this leak added to Reagan's war-mongering image. Accurate or not, the news coverage left the distinct impression that no administration ever had expressed such a sanguine view of potential nuclear conflict. The guidance document apparently transcended its immediate predecessor, Presidential Directive (PD) 59, both in its weapons requirements and in the ends to which those weapons could be the means.[20] It indicated that "United States nuclear capabilities must prevail even under the conditions of a long war . . . and be able to force the Soviet Union to seek earliest termination of hostilities on terms favorable to the United States";[21]

---

18. Weinberger presented this list of priorities along with a general defense of the program to the Senate Foreign Relations Committee during Reagan's first year in office. U.S. Senate, Committee on Foreign Relations, *Strategic Weapons Proposals: Part One*, 97th Cong., 1 sess., 1981, p. 193.

19. Richard Halloran, "Pentagon Draws Up First Strategy for Fighting a Long Nuclear War," pp. 1, 12. See also Christopher Paine, "Nuclear Combat," *Bulletin of the Atomic Scientists*, November 1982, pp. 5–12.

20. PD-59, President Carter's eleventh-hour hawkish revision of American declaratory nuclear strategy, signed in the summer of 1980, emphasized political and military targeting and the flexibility of nuclear options. Whereas PD-59 failed to match declared strategy to actual capabilities, the Pentagon's defense guidance statement was attached to an employment program. Jeffrey Richelson, "PD-59, NSDD-13 and the Reagan Strategic Modernization Program," *Journal of Strategic Studies* 6 (June 1983): 128–46. Richelson discusses the continuities and differences between PD-59 and Reagan's strategy.

21. Quoted in Richard Halloran, "Weinberger Confirms New Strategy on Atom War," *New York Times*, 4 June 1982, p. 10. For an analytic predecessor of Weinberger's strategy see Colin S. Gray and Keith Payne, "Victory Is Possible," *Foreign Policy* (Summer 1980): 14–27. With such articles in influential journals, Colin Gray breathed new

hence the controversy over fighting nuclear war. The continuum the document envisions between conventional and nuclear war, assumes a variety of uses for nuclear weapons, including the interdiction of Soviet conventional forces, the destruction of command structures and industrial facilities, and attacks on Soviet allies. Originally an internal Pentagon document, the defense guidance statement was later codified as National Security Decision Directive (NSDD) 13 and signed by Reagan.[22]

The foundation on which the Reagan doctrine was to rest was the vast material requirements of the five-year strategic weapons program covering each leg of the triad: the MX ICBM, the B-1 bomber and the cruise missile, and the Trident submarine and missile projects. Although most of these programs originated in previous administrations, Reagan planned to accelerate their development and production. The mothballed B-1 program was to be revived, and among other things, "black budget" research and development of the Stealth bomber would increase dramatically. Spending was to increase for command, control, and communications for protracted nuclear war or, at least, for the credible threat thereof. Furthermore, according to the administration, this delicate equation of infinite deterrence—a very different equation from MAD—perhaps required strategic defense against Soviet nuclear forces. The exacting doctrine had a place for every conceivable program.

Few, however, could have guessed that on 23 March 1983, President Reagan would announce that he was "directing a comprehensive and intensive effort to define a long-term research and development program to begin to achieve our ultimate goal of eliminating the threat posed by strategic nuclear missiles."[23] Two years into his presidency, Reagan inaugurated Star Wars, the most distinctive and innovative piece of the buildup. Within a year, the Pentagon had reorganized various existing ballistic missile defense programs under the central control of the Strategic Defense Initiative (SDI) Organization. Beginning with the FY 1985 defense budget, the funding for strategic defense rose rapidly in response to the planned five-

---

intellectual life into the war-fighting argument. He was subsequently appointed by Reagan as an advisor to the Arms Control and Disarmament Agency and the State Department.

22. Robert Scheer, "Pentagon Plan Aims at Victory in Nuclear War," *Los Angeles Times*, 15 August 1982, pp. 1, 12. Richelson, in "PD-59," identifies the plan Scheer reported as NSDD-13.

23. Text reprinted in *Congressional Quarterly Weekly Report*, 25 March 1983, p. 629.

year and $26 billion program. Unlike Reagan's other military policy initiatives, SDI represented not just a change in degree but in kind as well. Earnest pursuit of effective strategic defense had been eliminated from the U.S.-USSR geopolitical chessboard in 1972 with the ABM treaty; the SDI put it once again at the center of superpower politics.

In substance SDI was purely a research program with the goal of providing the "basis for an informed decision in the early 1990s on whether or not to develop and deploy a strategic defense system for the United States and its allies."[24] The research was divided into five areas: (1) surveillance, acquisition, and tracking; (2) directed-energy weapons; (3) kinetic-energy weapons; (4) battle management and systems analysis; and (5) support programs such as heavy-payload launch. The money to support this research agenda was spread widely among the three major military services, industries, and universities. Although the justifications for this extensive effort began to proliferate once the president offered Americans his vision of a world in which nuclear weapons were rendered impotent and obsolete, the ultimate goal of total defense remained the president's stated objective. Within a few years, SDI became the top priority of the Reagan military agenda and one of the largest programs in the defense budget.

The emphasis the White House, the press, and public interest groups were putting on nuclear weapons distracted attention from the conventional half of the Reagan buildup. The defense guidance statement of 1982 had devoted considerable attention to this potentially most important and most expensive half of the Reagan defense program. It discussed in detail scenarios of long-term conventional conflict with the Soviet Union and its allies on multiple fronts, the United States and its allies deliberately broadening the scope of the conflict by attacking vulnerable points in the Soviet sphere of influence. Such an approach came to be known as horizontal escalation or simultaneity. In essence, the United States and its allies might retaliate against Warsaw Pact aggression with attacks far removed from the immediate locus of conflict, which "should be launched against territory or assets that are of an importance to him compara-

---

24. U.S. Department of Defense, *Report of the Secretary of Defense Caspar W. Weinberger to the Congress on the FY 1988/FY 1989 Budget and FY 1988–1992 Defense Programs*, 12 January 1987 (hereafter cited as *FY 1988 Annual Report*), p. 281.

ble to the ones he is attacking."²⁵ This strategy would require great mobility, and rapid projection of U.S. military force could best be provided by the navy, along with the marines they transport. Accordingly, as Weinberger noted in 1982, "The most significant force expansion proposed by the Administration centers on the Navy, particularly those components of it that have offensive missions."²⁶ Not surprisingly, Secretary of the Navy John Lehman was among the most vocal exponents of the horizontal escalation strategy.

The administration's most important early initiative in conventional weaponry was the 600-ship navy (when Carter left office the navy had 479 ships). Supporters of naval expansion cited, in addition to the demands of horizontal escalation, the emergence of a Soviet blue-water navy. The most expensive and controversial part of the expansion was the planned addition of three aircraft carrier groups, which would raise the number from twelve to fifteen. To justify these multibillion-dollar armadas, Navy secretary Lehman and others pushed the idea of naval forward defense in the event of Warsaw Pact aggression in Europe.²⁷ Forward defense would move three or four carrier groups toward the Norwegian coast to intercept the Soviet navy, especially its submarines, before it can reach open

---

25. U.S. Department of Defense, *Report of the Secretary of Defense Caspar W. Weinberger to the Congress on the FY 1983 Budget, FY 1984 Authorization Request and FY 1983–1987 Defense Programs*, 8 February 1982 (hereafter cited as *FY 1983 Annual Report*), p. 16. In this document the secretary gives a more complete exposition on simultaneity and horizontal escalation: "The strategy we have been developing seeks to defend Alliance interests in such other regions [he has mentioned the Persian Gulf earlier]. For the region of the Persian Gulf, in particular, our strategy is based on the concept that the prospect of combat with the U.S. and other friendly forces, coupled with the prospect that we might carry the war to other arenas, is the most effective deterrent to Soviet aggression. This strategy, thus, has two dimensions. First, we must have a capability rapidly to deploy enough force to hold key positions, and we must be able to interdict and blunt a Soviet attack. It is the purpose of this capability to convince enemy planners that they cannot count on seizing control of a vital area before our forces are in place, and that they cannot therefore confront us with an accomplished fact which would deter our intervention. Second, this strategy recognizes that we have options for fighting on other fronts and for building up allied strength that would lead to consequences unacceptable to the Soviet Union," p. 14.

26. *FY 1983 Annual Report*, p. 30. For an overview of the navy's fortunes under Reagan and Lehman see "The United States Navy's on the Crest of a Wave," *Economist*, 19 April 1986, pp. 57–58. For an overview of Lehman's arguments and testimony see his "The Six-Hundred–Ship Navy," *Defense* (January/February 1986): 14–21.

27. H. C. Mustin, "The Role of the Navy and Marines in the Norwegian Sea," *Naval War College Review* 39, (March/April 1986): 2–6. Mustin was vice admiral and commander of Striking Fleet Atlantic. See also Robert S. Wood and John Hanley, Jr., "The Maritime Role in the North Atlantic," ibid. 38 (November/December 1985): 5–18.

sea, to protect Norway and its military assets from Soviet air forces, and to tie up Soviet forces that could be used on the central front. Whatever the justifications, the navy quickly embarked on its largest peacetime shipbuilding program.

Another important element in the materialization of the simultaneity doctrine was the Rapid Deployment Force (RDF), first planned for under the Carter administration.[28] In 1977 the RDF was conceived of as a small, mobile force for intervention in small conflicts; under Reagan, it quickly expanded to become a multiservice boondoggle. Originally intended to draw upon existing units from the various services, the RDF was assigned in the defense guidance statement of 1982 up to five army divisions (up from 3⅓ under Carter), two marine divisions (up from 1⅓), ten tactical fighter wings from the air force (up from 7), plus two B-52 wings and three naval aircraft carrier groups, showing generosity to each service.[29] The RDF, upgraded to U.S. Central Command (of South West Asia) in January 1983, thus went from about 220,000 troops to potentially 440,000 and was given geographic command of the Persian Gulf region.[30] The RDF became a central justification for airlift, sealift, and aircraft carrier procurement and for the Reagan administration's planned increase in the number of army divisions.

Nor did the Reagan administration ignore the army's principal domain, the European central front. The Pentagon accelerated old conventional programs and initiated new ones that embodied the high-technology and capital-intensive character of his other initiatives. This high-tech bias not only pervaded more traditional conventional weaponry such as tanks and helicopters but, more important, was at the center of what could be called the conventional defense initiative: the AirLand Battle doctrine and the Assault Breaker program.

Beginning with his first budget, Reagan accelerated the development and production of advanced-technology tanks (the M-1), fighting vehicles (the Bradley), air defense systems (DIVAD), and attack helicopters (the Apache AH-64) and the attendant weaponry and

---

28. In 1977, Carter signed PD-18, which called for the creation of a mobile strike force for non-NATO operations. Michael T. Klare, in "Army in Search of a War," *Progressive*, February 1981, pp. 18–23, provides a brief history of the RDF under Carter; for an apolitical and technical overview of the planning history for contingency forces from the 1960s to the 1980s see Robert P. Haffa, *The Half War* (Boulder, Colo., 1984).

29. Halloran, "Pentagon," p. 12.

30. For figures on these changes in the size see John D. Mayer, *Rapid Deployment Forces* (Washington, D.C., February 1983).

munitions. The FY 1983 budget claimed increases in investment over Carter's programs of 29 percent for tanks, 34 percent for fighting vehicles, and 25 percent for attack helicopters.[31] The army also began to invest more resources into high-technology light divisions. Each program complemented the army's new *Field Manual of Operations*, FM 100-5, adopted in 1982.[32] The manual puts greater stress on maneuver and mobility of defensive operations. Each system depends on advanced electronics and computers, much like spacecraft or jet fighters.[33]

The true conventional defense initiative stems from the central innovation of FM 100-5: the doctrine of AirLand Battle. The AirLand strategy calls for "deep strikes" by air- and ground-based conventional weapons against reserve Warsaw Pact forces to disrupt any attack.[34] At the heart of the five-year and ten billion–dollar effort are the development and deployment of various "smart" weapons or precision-guided munitions, identified by such acronyms as JTACMS, STAFF, and SADARM. Much of this kind of work is done in the Assault Breaker program jointly run by the army and the air force since 1982.

In general, U.S. policy has relied on qualitative superiority to offset any quantitative advantages in favor of the Warsaw Pact. Despite the numerous problems and inefficiencies associated with the more "conventional" high-tech systems, to say nothing of the unproven technologies at the heart of the conventional defense initiative, the qualitative superiority argument was used to support the Assault Breaker program.[35] As with the preceding Reagan military initiatives, the emphasis on capital-intensive, advanced technology is evident.

The combination of rapid increases and the character of the benefiting military programs had a profound effect on the distribu-

31. *FY 1983 Annual Report*, p. 29.
32. For a military analysis of the evolution of FM 100-5 see John L. Romjue, *From Active Defense to AirLand Battle* (Fort Monroe, Va., June 1984).
33. Each system experienced problems with cost overruns, slow production rates, dubious performance, and safety.
34. The seminal piece on deep strikes is Donn Starry's "Extending the Battlefield," *Military Review* 61 (March 1981): 31–50. For an early and good overview of what AirLand meant for NATO's defense see Deborah Shapley, "The Army's New Fighting Doctrine," *New York Times Magazine*, 28 November 1982, pp. 36–42+.
35. For example, a favorable report done for two House subcommittees argues that such technologies act as "force multipliers," which obviate numerical competition. *Improved Conventional Force Capability*, staff study prepared for the subcommittees on Research and Development and on the Military Nuclear Systems of the House Armed Services Committee, 98th Cong., 1st sess., 1984, p. 5.

tive impact of the defense budget. Not all categories and services benefited equally from the general buildup. The new doctrines and programs were used to justify a dramatic shift from operational expenditures to investment in research and development and procurement, which created a highly capital-intensive military budget. The percentage of the defense budget devoted to nuclear arms increased from 7 to 10 percent, and actual spending on strategic programs more than doubled, from about $13 billion in 1981 to about $28 billion in 1985. Next, the general category of procurement increased greatly from nearly 25 percent of the defense budget in 1980 to 33 percent in 1986. Actual spending on procurement nearly tripled, from about $35 billion in 1980 to $93 billion in 1986, making procurement the single largest category in the military budget. Research and development (R&D) grew from 9.5 percent to 12 percent, an increase of over 31 percent, almost a tripling of actual spending, from $13.6 billion in 1980 to $33.6 billion in 1986. Some of the increase in research and development (R&D) and procurement was hidden in Reagan's expansion of the so-called black budget, which consisted of named and anonymous research and procurement programs for which no details are given. Reagan increased the portion of the Pentagon budget that is black, or secret, by a substantial 300 percent.[36] Because Congress generally respects this secrecy, little scrutiny is given to these programs. They are, consequently, difficult to monitor and difficult to cut.[37]

Relatively neglected was the operational side of the defense budget, including personnel compensation, stockpiles, and maintenance. In actual spending, all categories increased during the buildup, even in real terms. But operating expenditures, although historically over 60 percent of defense spending, typically have had a lower priority in military budgets; there has seemed never to be enough money for adequate compensation, training, and stockpiles of ammunition and supplies. Given Reagan's emphasis on procurement of new weapons and increases in personnel, operating expenditures would have had to increase proportionately to investment to

36. David C. Morrison, "Pentagon's Top Secret Black Budget Skyrockets," *National Journal*, 1 March 1986, pp. 492–98. The black budget increased from about $5.5 billion in 1981 to more than $22 billion in the 1987 request, and incorporated close to 20 percent of military R&D funding. For a detailed study of the black budget, see Tim Weiner, *Blank Check* (New York, 1990).

37. For example, in 1989 the first test B-2, or Stealth, bomber was unveiled, with the revelation that more than $22 billion had already been spent just on development under this black program.

Table 5. Growth of the capital-intensive defense budget (in billions of current dollars)

	1980	1983	1986	% increase 1980 to 1986
Investment				
Procurement	35.3 (24.8%)	80.4	92.5 (32.9%)	
RDT&E[a]	13.6 (9.5%)	22.8	33.6 (11.9%)	
Construction	2.3 (1.6%)	4.5	5.3 (1.9%)	
Total investment	51.2 (35.5%)	107.7	131.4 (46.7%)	157
Operations				
Personnel	43.0 (30.2%)	61.7	67.8 (24.1%)	
Operations and maintenance	46.4 (32.5%)	66.5	74.9 (26.6%)	
Total operations	89.4 (62.7%)	128.2	142.7 (50.7%)	60

Sources: Figures calculated from *Department of Defense, FY 1986 Annual Report* and *FY 1988 Annual Report*.

Note: In parentheses are the percentage of the total military budget represented by the preceding dollar figure; numbers may not add to 100 because of rounding and minor appropriations categories not included.

[a]Research, development, testing, and evaluation.

keep even normally behind. Such was not the case. Overall, total operating expenditures dropped from almost 63 percent of the military budget in 1980 to slightly more than 50 percent in 1986 (Table 5). For example, personnel expenditures decreased from 30.2 percent to 24.1 percent, and operations and maintenance from 32.5 percent to 26.6 percent. At the same time, overall investment increased from 35.5 percent to 46.7 percent. This represents a nearly 160 percent increase in investment funding while operating funds increased by only 60 percent. The gap between investment and operations grows even wider when Department of Energy (DOE) military programs—mostly nuclear warhead production, almost pure R&D and procurement—are included; DOE military programs grew 81 percent in real terms from $3 billion in 1980 to $7.5 billion in 1987.

Finally, the character of military capital investment changed under Reagan's guidance. Procurement can take place over a short time (munitions, small arms, and other simpler weapons and machines) or over many years (aircraft, ships, missiles, and other advanced weapons and weapons platforms). The Reagan programs expanded the portion of the military budgets devoted to multiple-year procurement contracts, creating an enormous backlog of budget authority that would come due as outlays in future years.[38] Such obligations add to the "uncontrollable" part of the military budget. A projection by the Congressional Budget Office (CBO) in 1987 illustrated this phenomenon: assuming a zero real growth rate in budget authority, the CBO calculated that outlays from FY 1987 to 1992 would total $1.8 trillion compared with $1.3 trillion from FY 1981 through 1986, an increase of 38 percent.[39] In fact, although small but real decreases in spending authority from FY 1986 onward changed the total dollar figures, the CBO forecast has proved correct. As we shall see in Chapter 8, total outlays from the second half of the decade and 1990 significantly exceeded outlays through 1985. Just as the Democrats institutionalized higher levels of social spending during the New Deal and the 1960s, the Republicans, under Reagan, engineered a significant commitment to greater military spending.

Military Keynesianism and a Military Industrial Policy

This investment-intensive, advanced-technology military buildup had two broad economic purposes and consequences beneficial to the institutionalization of the Reagan regime: it functioned as a Keynesian stimulus to the economy and as an industrial policy. Despite his public disdain of big government and governmental intervention in the private sector, Reagan increased both through his military programs. The economic stimulation that his public philosophy would not permit through explicit economic programs was accomplished through his national security program. Consequently, while Reagan's ideological appeals garnered support from many who did not benefit economically from the military buildup, the investment in defense generated commitment to his regime on the part of its economic beneficiaries in labor, management, and finance. Although the budgetary contradictions resulting from domestic cutbacks, tax reform, defense spending, and the deficit weakened

---

38. Center on Budget and Policy Priorities, Defense Budget Project, *The Fiscal 1986 Defense Budget: The Weapons Buildup Continues* (Washington, D.C., 1985); Seymour J. Deitchman, "Weapons, Platforms, and the New Armed Services," *Issues in Science and Technology* 1 (Spring 1985): 83–99.
39. Michael Ganley, "Defense Buildup Will Slow, But Outlays Will Increase 38% over Next Five Years," *Armed Forces Journal International* (March 1987): 8.

public support for a continued binge of defense spending, the administration retained the support of large segments of industry and the public for a continuation of the dominance of the warfare state over the welfare state.

During his first administration, Reagan put military Keynesian economics to work in two interrelated ways. First, when the severe recession of 1982 set in, a heavy influx of military spending and investment bolstered the economy. Second, once military-related production created some of the few bright spots in an otherwise dismal economic picture, the administration used that fact to gain support for further increases in military spending.

The FY 1982 budget produced 12 percent real growth in military budget authority and an 8.1 percent real increase in military outlays, or actual spending, for that year. This capital infusion boosted certain industries in troubled times; military spending accounted in particular for the overwhelming portion of durable goods orders during the recession. By the end of 1982 the Commerce Department could report that new orders for durable goods rose a record 12 percent in December. Out of that total increase of $8.5 billion in orders, defense goods accounted for $6.9 billion, or about 75 percent, while other orders only contributed $1.6 billion.[40] This increase foreshadowed increased production and lower unemployment in the coming months—in short, a recovery from the recession. While from 1981 to 1984, industrial production grew by only 3 percent overall, "production in the defense and space industries . . . soared by 9.5%."[41] As employment decreased in smokestack industries, and as Reaganomics and the recession took effect, employment increased in defense-dependent sectors. For example, employment in the aerospace industry, which had reached a low of 893,000 in 1977, soared to over 1,220,000 by September 1984.[42] The Defense Department, moreover, hired a signifcant number of new military and civilian employees. Whereas both active military personnel and direct-hire civilian Pentagon employment decreased under Carter, Reagan increased each by over 100,000 for a total increase of 230,000. As a result, Department of Defense (DOD) direct-hire and industrial em-

---

40. John M. Berry and David Hoffman, "Inflation in 1982 was 3.9 Percent, Least in a Decade," *Washington Post*, 22 January 1982, pp. 1, 5.
41. "Pentagon Spending Is the Economy's Biggest Gun," *Business Week*, 21 October 1985, pp. 60–61.
42. Industry figures cited by Mary H. Cooper, "The Defense Economy," *Editorial Research Reports*, 17 May 1985, pp. 359–76.

ployment went from 4.7 percent of the national work force in 1980 to 5.5 percent in 1986.[43] As for the overall effect on employment of the Reagan buildup, Nariman Behravesh of Wharton Econometrics says, "Defense spending increases probably provided the greatest momentum to growth in recent years. About 15% to 20% of the employment gains we've seen in the past three years [1983–85] are directly or indirectly due to defense spending."[44]

The implications and potential of such statistics were not lost on the administration, and they and their allies set about making sure that Congress was also aware of them. In his annual report to Congress on the 1984 budget, delivered in February 1983, Weinberger offered evidence about the positive effect of military spending on employment, and he also warned against using the military budget to cut the deficit inasmuch as cuts in programs generally have only a small yearly impact. Trimming defense, he maintained, would do less to cut the deficit than it would to reduce revenues, because "a large part of DOD expenditures become income to firms and individuals, some of which comes back to the government in taxes; and the ripple effects from defense spending tend to stimulate growth in GNP, which also increases total tax revenues."[45] As the 1984 election approached, Weinberger put the case more bluntly: "The defense budget doesn't contribute to the deficit to the extent that other types of spending do. Defense spending increases tax revenues and produces a great many jobs. Indeed, it's one of the leading factors in the economic recovery."[46]

Despite his rhetorical assault on big government, Reagan used and depended on Keynesian economics, based on military spending, for the partial economic recovery in late 1982 and 1983. He could then argue for further increases in defense spending. Redistribution of billions of tax dollars, no longer justifiable for social programs, became a political and economic necessity under the banner of national security.

In July 1986, LTV Corporation, a conglomerate involved in steel, energy, and aerospace and defense production, filed for bankruptcy

43. *FY 1988 Annual Report*, p. 327, table 4.
44. Quoted in *Business Week*, 21 October 1985, p. 60.
45. U.S. Department of Defense, *Report of the Secretary of Defense Caspar W. Weinberger to the Congress on the FY 1984 Budget, the FY 1985 Authorization Request, and the FY 1984–1988 Defense Programs*, 1 February 1983 (hereafter cited as *FY 1984 Annual Report*), p. 67.
46. Interview with *U.S. News and World Report*, 9 July 1984, p. 29.

Table 6. Sales and income for the LTV Corporation, 1981–85 (in millions of dollars)

	1981	1982	1983	1984	1985
Sales					
Steel	4,786	3,040	2,935	4,521	5,375
Energy	2,081	1,203	501	647	592
Aerospace and defense	797	777	1,142	1,953	2,259
Operating income or loss					
Steel	336	−299	−200	−217	−227
Energy	220	35	−57	−73	−26
Aerospace and defense	39	45	67	125	164
Net	595	−219	−190	−165	−89

Sources: Jonathan P. Hicks, "Merger Policy Fails at LTV," *New York Times*, 18 July 1986, p. D1.

protection in the largest such procedure in U.S. history at the time. Its financial statistics (Table 6) tell a graphic tale of corporate decline. Without its military contracts, LTV would have had to file for bankruptcy sooner. At the time of the bankruptcy it looked to unload its steel operations and concentrate on defense production. As a senior vice-president of LTV put it, "The crown jewel of the company happens to be the aerospace and defense company, and we want it to be continued and enlarged."[47] LTV's plight seems a good illustration of the economic trends Reagan's policies fostered: declining traditional industries and reliance on defense-funded high-tech operations.

Just as Reagan objected to Keynesian economics, he had a similar distaste for industrial policy. Nonetheless, the quantity and quality of the defense spending that produced Keynesian effects in the economy also implemented a de facto industrial policy that rewarded certain economic sectors.[48] The winners were high-tech

---

47. Robert L. Simison and Cynthia F. Mitchell, "LTV's Strategy May Be Gaining Support," *Wall Street Journal*, 22 July 1986, p. 6.
48. For a similar development of this point, see Ann Markusen, "The Militarized Economy," *World Policy Journal* 3 (Summer 1986): 495–516; she includes the rapid rise in foreign military sales facilitated by the Reagan regime. Recognizing the influence of military spending, some have advocated making the transition to an explicit and comprehensive industrial policy through the defense market, because of the existing high level of governmental involvement and the long-term growth potential of its primary industries; see Jacques S. Gansler, "Defense," *Challenge* 26 (January/February 1984):

industries involved in electronics, aviation, and shipbuilding. Pentagon policy targets growth industries and provides them with research and development funds and often with subsidized plant facilities. The DOD acts as a principal consumer for their products and even provides job retraining and promotes trade for its target sectors. Nonmilitary aspects of Reaganomics augmented the effects of military spending by hurting older industries while helping high-tech sectors.[49]

The quantity of military capital the Reagan administration directed into research, development, and procurement of advanced technologies vastly exceeded any explicit plans for a national industrial policy proposed by Gary Hart or anyone else in the early 1980s. Hart's plans, in fact, involved many virtually cost-free elements, including governmental studies and GATT negotiations; the budgetary items included job retraining ($103–$717 million per year), loans for modernization (perhaps one appropriation of about $5 billion), and more civilian R&D funds for high-tech ventures. The Industrial Competitiveness Act, sponsored unsuccessfully by Representative John LaFalce (D-N.Y.) in 1984, called for, among other things, the creation of a bank for industrial competitiveness to provide $8.5 billion in loans for modernization, a figure that pales before the over $25 billion the Pentagon spent (not loaned) for industrial research and development in 1986. McDonnell-Douglas alone received over $7 billion in contracts in 1984.

The real impact of defense spending is not measured in absolute terms, however, but in the increased investment into procurement and research. Whereas Carter's real increases in procurement and R&D were barely above 1976 levels (a cumulative total of about 20 billion in constant dollars from 1977 to 1980), Reagan's were unprecedented. In R&D alone he accumulated a total for FY 1982–85 of $30 billion above 1981 levels. In the late 1970s about 50 percent of federal R&D was military; the Reagan budgets increased that figure to about 70 percent. In 1985, over one-third of the United States's total R&D investment was military, whereas in Germany the military consumed about 4 percent, and in Japan about 1 percent, of R&D.[50]

58–61.
49. Robert Reich, "Reagan's Hidden 'Industrial Policy,'" *New York Times*, 4 August 1985, sec. 3, p. 3.
50. Robert Reich, "High Tech, a Subsidiary of Pentagon Inc.," *New York Times*, 29

Federal military R&D accounted for over 60 percent of all high-tech R&D funds for the country by 1985. As for procurement, the total real increases made under Reagan approach $120 billion for FY 1982–85. While not all of this large influx of capital went toward high-tech industries, much of it was concentrated in electronics and aerospace. Even the M-1 tank's price tag is in large measure due to advanced electronics.

Soon after the Reagan buildup began, the business community took note of the direction in which it was leading industry. *Fortune* noted that military spending "is bringing important changes to U.S. industry. As MIT economist Lester C. Thurow observes, 'The Department of Defense conducts an industrial policy, though it is a very narrowly focused one.' Over the years, it has helped strengthen the industries that are the U.S.'s most potent international competitors, including aerospace, computers, scientific instruments, and communications equipment."[51] Electronics were pegged as one of the fastest growing markets in late 1982, with great potential for stock values.[52] Texas Instruments, an advanced electronics firm, relied increasingly on military contracts to offset losses in its commercial markets.[53] Within weeks after Reagan's Star Wars speech in March of 1983, stocks in companies with existing strategic defense contracts or SDI potential rose significantly in price.[54] Defense purchases were projected to produce nearly half of the growth in the aerospace industry and 20 percent of the growth in electronics through 1987, with similar impetus expected in the fabricated metals and machine tool industries.[55]

Traditional heavy industry reached out to advanced-technology operations to get in step with the perceived future of the economy. In 1985, General Motors acquired Hughes aircraft, a long-time leading defense contractor and near the top in SDI funding. Thus a giant of the smokestack industries sought to diversify and build for the future by absorbing a large high-tech firm. Chrysler, which was

---

May 1985, p. 23.

51. Bruce Steinberg, "The Military Boost to Industry," *Fortune*, 30 April 1984, pp. 42–48.

52. *Business Week*, 20 September 1982, p. 80.

53. Thomas C. Hayes, "Texas Instruments Is Back in the Black," *New York Times*, 26 July 1986, p. 35.

54. "Beaming in on Star Wars Stocks," *Business Week*, 25 April 1983, p. 120.

55. Steinberg, "Military Boost to Industry."

forced to sell its profitable M-1 tank operation to General Dynamics in 1982, had to abandon its attempt to purchase a high-tech company in the $2 billion range because none was available for a friendly takeover.[56]

The Reagan Revolution, the New Great Equation, and the Democratic Dilemma

While significant, these macroeconomic impacts of the buildup did not compare to what was transpiring at the political epicenter on Capitol Hill. The Reagan buildup was a key component of the larger Reagan revolution—the most significant alteration of governmental programs and priorities since the New Deal and World War II. As Samuel Huntington has argued, the formation of defense policy and programs takes place within the framework of the "Great Equation" of the postwar American political economy.[57] Administrations have had to balance four competing priorities: (1) domestic programs, or "welfare"; (2) foreign and defense policies, or "security"; (3) tax limitations, or the "private sector"; and (4) balanced budgets, or "fiscal integrity." While each administration used slightly different mathematics, for the first two decades of this era a growing economy made the balancing act less difficult. Truman sought to limit both domestic and defense spending but then budgeted the massive defense-spending increases for the Korean War. Eisenhower wanted to limit taxes and the deficit by keeping domestic and defense spending down. With Kennedy and Vietnam began the long-term rise of both domestic and defense spending at the expense of lower taxes and a balanced budget. Johnson's attempt simultaneously to produce the Great Society and to win the war in Vietnam brought guns and butter into painful conflict. As defense spending leveled-off and began to decline in real terms after the Vietnam peak in 1968–69, domestic spending rose as a percentage of the budget and GNP. The oil shocks and stagflation of the 1970s, however, exacerbated programmatic and fiscal tradeoffs, and federal budget deficits grew significantly.

Reagan seized on widespread discontent with the economy, waste in governmental programs, and taxes. The Reagan Great Equation

---

56. *New York Times*, 24 July 1986, p. D7.
57. Samuel P. Huntington, *The Common Defense* (New York, 1961), pp. 197–98.

reversed much of the mathematics of the Democratic welfare state that had dominated the federal budget over the previous three decades. As Martin Shefter and Benjamin Ginsberg observed at the time, the "central element of the Reagan strategy is to shift the focus of national political attention from the expenditure to the revenue side of the federal budget."[58] Reagan gave political substance to public discontent by mobilizing popular opinion against the federal tax system and the domestic governmental programs that produced high taxes. He also altered the political orientation of those who had seen themselves as consumers of governmental programs; by stigmatizing deficit spending and most forms of social spending, he shifted their political attention to taxes and governmental expenditures. Reagan applied this anti–big government ideology, however, to only one half of the output side of the great equation: domestic programs. Defense programs remained relatively sacrosanct as the truly legitimate output of the federal government, the one necessary national public good. The Great Equation of the Reagan era, therefore, was a combination of lower taxes, a massive defense buildup, cuts in domestic spending, and resulting large deficits.

At the same time, then, that he was engineering the defense buildup, Reagan and his congressional allies were legislating decreases in domestic programs and the largest tax cuts in recent American history. The administration's tax cuts, which emerged from Congress as the Economic Recovery Act of 1981, projected a $737 billion loss in federal tax revenues over the next five years. The tax legislation regressively lowered personal income tax rates and top rates on unearned income and limited the inheritance tax. It also cut corporate income taxes and capital gains rates. Although amendments to this plan in 1983 and 1984 slowed and decreased the revenue losses, taxes decreased from 20.1 percent of GNP to 18.6 percent from 1981 to 1985.[59] The only tax that rose in real terms was the regressive social security tax.

The only way the Great Equation could stay in balance was for the domestic cuts to be large enough to balance both the defense in-

---

58. Martin Shefter and Benjamin Ginsberg, "Institutionalizing the Reagan Regime" (Paper presented at the annual meeting of the American Political Science Association, New Orleans, 1985), p. 2. Ginsberg and Shefter provide a cogent and complete analysis of the Reagan political and economic strategy in their *Politics by Other Means: Institutional Conflict and the Declining Significance of Elections in America* (New York, 1990).
59. John E. Schwarz, *America's Hidden Success* (New York, 1988), p. 150.

creases and the revenue losses. For fiscal 1982, Reagan negotiated with Congress for about $35 billion in cuts in hundreds of domestic programs.[60] Victims included the Department of Housing and Urban Development, whose budget would plummet from $35.8 billion in 1980 to $14.7 billion in 1987.[61] Many programs were cut or trimmed as federal-to-state grants were consolidated. Job-training and employment programs went from $10.3 billion in 1980 to $4.6 billion in 1984. Community and regional development grants were slashed. Stricter eligibility requirements slowed spending on Medicare and Aid to Families with Dependent Children. Many educational programs were reduced. Even energy-supply and conservation programs shrank significantly. Despite these attempt at massive cuts in domestic spending, massive deficits were the nearly inevitable consequence of the new equation. The budget deficit skyrocketed from $78.9 billion in 1981, to $127.9 billion in 1982, to $207.8 billion in 1983, eventually doubling the total federal debt by 1986 (to $2,130 billion from $1,003.9 billion in 1981). As a result, interest payments paid by the federal government increased from $52.5 billion in 1980 to $136 billion in 1986.

Despite congressional resistance to both the rate of increase in the military budget and the extent of cuts in domestic programs, Reagan engineered a dramatic revision of the federal budget. Discretionary domestic programs fell from 5.7 percent of GNP in 1981 to 3.7 percent by 1987, while defense spending rose from 5.3 to 6.4 percent. In budgetary terms, this change is more striking: from FY 1981 to 1987 discretionary spending on domestic programs *decreased* by 21 percent in real terms while defense outlays *increased* by 45 percent.[62]

Only the political power of national security issues allowed Reagan to effect such a transfer of resources from domestic to defense programs while allowing deficits to skyrocket. Although defense spending would eventually level off and public enthusiasm for the buildup would wane, Reagan had raised the American military commitment to a new and very expensive plateau. As long as the Soviet threat remained credible, significant cuts were politically impossible. The close to $300 billion annual military budget had become the new standard; anything less would be considered unilateral disarma-

---

60. John Cranford, *Budgeting for America*, 2d ed. (Washington, D.C., 1989), pp. 136–37.

61. These and the following figures come from Office of Management and Budget data.

62. The figures come from the Congressional Budget Office.

ment. Moreover, this budget shift would prove difficult to reverse, regardless of perceived need, insofar as the capital-intensive nature of the buildup made quick and easy cuts in the military budget more difficult. At the same time, the tax cuts became the entitlement programs of members of middle- and upper-class socioeconomic groups and, therefore, also politically dangerous to rescind or revise. The deficit became a hot political issue, with Reagan and Democrats in Congress blaming one another for its malignant growth. The problem for the Democrats, however, was that attention to the deficit only made increases in domestic programs all the more improbable. Consequently, the first and principal political victims of the Reagan buildup were not the Soviets but the Democrats. From the beginning, Reagan's Great Equation effectively neutralized the traditional Democratic agenda. The sweeping impact of just the first year of the Reagan presidency left Democrats stymied and strategically adrift, well before the full effects could be known. Yet even as Democratic politicians pondered the onslaught and began the search for a response, the first counterattack against the Reagan administration was taking shape. The largest of peacetime military buildups was soon to be contested by the largest peacetime peace movement.

3

The Rise of the Nuclear Weapons Peace Movement

As conservative political forces began to mobilize around issues of foreign and defense policy in the late 1970s, increasing numbers of activists of the liberal-left were redirecting their attention and energies toward the problems of nuclear weapons, strategy, and potential warfare. Just one year after Reagan's election, this activity burgeoned into the largest peacetime peace movement in the nation's history. The nuclear weapons peace movement became the dominant liberal political movement of the 1980s. In some ways, it stands as the most remarkable in the series of left-liberal social movements of the past three decades. When circumstances combined to kill détente and resurrect public concern about nuclear weapons and superpower conflict, the contemporary peace movement took shape. Old groups grew and new groups formed. Reagan's election and policies greatly accelerated this process, which in less than two years produced the New York City rally of 1982, perhaps the largest political demonstration in U.S. history. Never before had a movement so rapidly accrued allies in the mass media and so quickly affected public consciousness and public opinion.

The importance of the peace movement in the 1980s extends beyond its effect on American arms control and nuclear weapons policies. The political significance of the movement will not be discovered in a recounting of its efforts and achievements, its successes and failures. Viewed from the perspective of domestic political competition, the peace movement played another crucial role in the politics of the 1980s: it made issues of nuclear war and arms control into the political Achilles' heel of the Reagan administration. The peace movement and its issues thus became the principal weapon liberals and the Democratic party used to contest resurgent conservatism and the Reagan revolution.

Origins: Three Cycles in the American Peace Movement

1957–1963: The Movement at the Periphery

The origins of the 1980s American peace movement extend back to the Eisenhower-Kennedy years. After a decade of quietude following the onset of the nuclear age, the very existence and potential consequences of atomic weapons became a source of public concern and political debate.[1] Atmospheric H-bomb tests prompted release of scientific evidence on the dangers of fallout. In 1955, members of the scientific community began to urge a limit or ban on testing. During his 1956 presidential campaign against Eisenhower, Adlai Stevenson turned the latent concern into a national issue with his proposal for a test ban. Although the issue ultimately backfired for Stevenson, it did raise public awareness; pressure grew for the administration to mitigate the testing problem.[2]

During the next few years, a new peace movement emerged, the first motivated by nuclear weapons and, for that matter, the first motivated by the dangers of a specific military technology.[3] The new movement grew from the expansion of three traditional constituents of the peace movement: middle- and upper-middle-class liberal reformers, religiously based organizations, and pacifist groups. In addition, scientists emerged as a distinct community in the peace movement. During these years, new groups from each sector organized and grew rapidly. SANE, soon to be the predominant middle-class organization, was founded in 1957. One of the more radical

---

1. The mostly unsuccessful efforts of peace groups and concerned scientists from 1945 to 1957 are covered by Lawrence S. Wittner, *Rebels against War* (New York, 1969), chaps. 6–8.
2. Robert A. Divine, *Blowing on the Wind* (New York, 1978).
3. For more on the peace movement in these years see Charles De Benedetti, *The Peace Reform in American History* (Bloomington, Ind., 1980), chap. 7; and Wittner, *Rebels against War*, chaps. 9–10.

organizations was the Committee for Non-Violent Action (1958). The first Pugwash conference of international scientists was held in 1957. Students organized the Student Peace Union in 1958. Most other existing peace groups, especially religious ones, experienced a rebirth of support and activism. Activities from lobbying to civil disobedience took place across the country in support of a test ban and disarmament.

After the testing issue failed to benefit Stevenson, neither political party would embrace the movement or its issues. In fact, much of the scant attention peace activists received was negative. For example, in 1960, Thomas Dodd, first-term Democratic senator from Connecticut, attacked SANE, charging that the organization was infiltrated by communists. His accusation divided and damaged the young organization.[4] Similar charges would be leveled against subsequent peace efforts, including the nuclear freeze mobilization in the 1980s.

The movement peaked well before it could become an influential force in national politics. In 1961, Kennedy and Congress established the Arms Control and Disarmament Agency. In 1963 came the partial test ban treaty, which restricted the testing of nuclear devices to underground. These measures seemed to be part of the administration's "opening to the left," after its initially hawkish cold war policies. Irrespective of Kennedy's death, the opening proved evanescent. Concern over nuclear weapons dissipated as peace activists confronted the war in Southeast Asia and joined the civil rights movement.[5]

1965–1975: The Antiwar Movement and the New Politics

With the exception of sporadic activism over antiballistic missiles in 1968 and 1969, nuclear weapons were all but ignored from 1963 to the late 1970s, despite the fact that this period saw the largest growth in the superpowers' strategic arsenals.[6] Gallup polls in 1959

4. Milton S. Katz, *Ban the Bomb* (New York, 1985).
5. For example, although the leadership of the mainstream peace movement was largely reluctant to address U.S. involvement in Vietnam, many people in the movement started to do so. At the 1963 annual Easter ban-the-bomb marches, some activists carried signs condemning U.S. actions in Vietnam. For this and more on the transition to Vietnam see Fred Halstead, *Out Now!* (New York, 1978).
6. Paul Boyer, "From Activism to Apathy," *Journal of American History* 70 (1984):

showed that 64 percent of Americans listed war (especially nuclear war) as the most important national problem. In 1965 only 16 percent said so. In 1969, 63 percent cited the Vietnam War, and a scant 2 percent named nuclear conflict. Nevertheless, these years developed the political skills and the social base that would constitute the nuclear peace movement of the 1980s. Demographic changes combined with events to produce a greatly expanded potential constituency for the next effort against nuclear weapons.

To a much greater degree than during the antitesting/ban-the-bomb era, the Vietnam years introduced the peace movement to coalition building and more radical forms of political action. Groups such as SANE, not without internal dissent, put nuclear weapons on the backburner and devoted most of their energy to the Vietnam War. They joined in the largely ad hoc coalitions that were the backbone of the antiwar movement.[7] The peace movement quickly became part of a larger alliance and a broader political agenda.

The civil rights movement and the Vietnam War combined to shape the political perceptions and motivations of a considerable segment of the baby boom generation. Central to this aggregation were students and academic communities, and from them sprang the New Left, those most radicalized by the twin experiences of racist oppression at home and abroad. Despite its disagreements with the New Left, the liberal core of the test ban/disarmament campaign was also radicalized to some extent by the war and domestic strife. Martin Luther King's call to end the war strengthened the links among liberals, the left, and black activists, so that, despite dissension and divisiveness, they were able to come together in tenuous alliance on the issues of the war, civil rights, and domestic priorities.[8]

821–44. In addition to the important distractions of Vietnam and civil rights, Boyer discusses other causes for the inattention to nuclear arms. The exception during these years was the stir over the proposed deployment of ABMs in 1968 and 1969. But this never became a national movement and was short-lived.

7. For histories of the antiwar movement with special emphasis on its organization and internal politics see Nancy Zaroulis and Gerald Sullivan, *Who Spoke Up? American Protest against the War in Vietnam, 1963–1975* (New York, 1984); and Halstead, *Out Now!*

8. De Benedetti, in *Peace Reform in American History*, pp. 171–80, discusses the ups and downs of this coalition in the mid-1960s. One example of the latter occurred at the September 1967 National Conference for a New Politics in Chicago, which was reportedly dominated by white middle-class liberals and characterized by difficult relations with the Black Caucus pressing for a more radical policy agenda; see Derek M. Mills, "From New Politics to Mass Catharsis," *War/Peace Report*, October 1967, pp. 8–9. For a more comprehensive discussion of the coalitional politics of the antiwar

Demographically, each constituent group of this alliance—white liberals, students, and politically active blacks—was growing rapidly in these years. The civil rights movement made black political power tangible through the emergence of the black vote. Baby boomers swelled the universities and colleges, producing a surge in the numbers of full-time students and the college educated, especially women. The increase in the educated population coincided with the expansion of the service, technical, and professional sectors of the economy, those areas of employment in the postindustrial society that encompass what has been called the New Class.[9] The New Class of predominately white middle- and upper-middle-class professionals has been associated with a disposition toward "post-materialist" values such as environmentalism and other liberal views.[10]

This combination of generational experiences and socioeconomic change spawned what has come to be known as the New Politics movement.[11] Vietnam and the civil rights movement combined with the growth of the professional sectors and the college student population to create its core constituency. The New Politics comprised the series of liberal issues and movements that flourished in the late 1960s and 1970s, including open government, antimilitarism, environmentalism, consumerism, and the women's movement.[12]

The national power of the coalition was manifest most dramati-

---

movement see Halstead, *Out Now*, in which he also discusses the New Politics Conference (pp. 316–21).

9. On the New Class, its origins, and role in postindustrial society see Daniel Bell, *The Coming of Post-Industrial Society* (New York, 1973); and Barbara Ehrenreich and John Ehrenreich, "The Professional-Managerial Class," in *Between Labor and Capital*, ed. Pat Walker (Boston, 1979). What perhaps best characterizes members of the New Class is their abundance of human capital, that is, advanced education and training.

10. Stephen Brint, "'New Class' and Cumulative Trend Explanations of the Liberal Political Attitudes of Professionals," *American Journal of Sociology* 10 (1984): 30–71; B. Bruce-Briggs, ed., *The New Class?* (New Brunswick, N.J., 1979).

11. As editor of *Nation* in 1962, Carey McWilliams wrote that "new arrivals" such as students, middle-class housewives, and blacks were united by the their desire for a "new politics." "Time for a New Politics," *Nation*, 26 May 1962, p. 466. For further analysis of the groups amenable to this movement see James A. Burkhart and Frank J. Kendrick, eds., *The New Politics* (Englewood Cliffs, N.J., 1971); for comprehensive history from its origins in the civil rights and antiwar efforts to its attempts to gain power in the Democratic party through 1972 see Larry J. Davis, *The Emerging Democratic Majority* (New York, 1974).

12. Many analysts commented on the new political values that underlay these issue movements; see, for example, Warren E. Miller and Teresa Levitan, *Leadership and Change* (Cambridge, Mass., 1976); on the relation of socioeconomic change to new political values see Ronald Inglehart, *The Silent Revolution* (Princeton, N.J., 1977).

cally during the 1972 presidential campaign, especially at the Democratic convention in Miami Beach. After the 1968 convention, the McGovern-Fraser Commission on Party Structure and Delegate Selection had produced eighteen guidelines for state selection of delegates for 1972. These rules had two major purposes and consequences. First, they opened or democratized the process of delegate selection, by reducing the control of party organizations. Second, they compelled increased representation of women, youths, and minorities. At the 1972 convention, 40 percent of the delegates were women (versus 13 percent in 1968); 21 percent were aged thirty or under (versus 2.6 percent); and blacks went from less than 6 percent to 15 percent in 1972. Correspondingly, the number of party officials and officeholders was greatly reduced. Led by George McGovern, the nascent coalition wrested control of the party from the New Deal party regulars.[13] The resulting platform, probably the most liberal in the nation's history, endorsed a guaranteed income, national health insurance, and progressive tax reform. The first order of business was an immediate and complete withdrawal of U.S. military forces from Southeast Asia. Control over issues of war and peace were to be returned to Congress and the people. Defense budget cuts would finance domestic programs. As the Democratic nominee, McGovern went on to campaign on these themes, concentrating on the war in Vietnam.

Although McGovern's ignominious defeat spurred the old Democratic regulars to return the party to its New Deal moorings, Watergate and the powerful appeal of New Politics issues gave further impetus to the liberal movement. Through the 1970s, New Politics movements dominated the political agenda.[14] Environmentalism, consumerism, and the open government and women's rights causes found legislative and judicial power through the Democratic majorities in Congress and the federal courts. Furthermore, the War Powers Resolution, amendments to the Arms Export Control Act, tighter controls over CIA activities, and termination of covert mili-

---

13. See Nelson W. Polsby, *Consequences of Party Reform* (New York, 1983); Byron E. Shafer, *Quiet Revolution* (New York, 1983).

14. See Benjamin Ginsberg and Martin Shefter, "The Setting," in *The Elections of 1984*, ed. Michael Nelson (Washington, D.C., 1985); Jo Freeman, ed., *Social Movements of the Sixties and Seventies* (New York, 1983); David Vogel, "The Public Interest Movement and the American Reform Tradition," *Political Science Quarterly* 95 (Winter 1980–81): 607–27. On the New Politics in Congress see Thomas P. Murphy, *The New Politics Congress* (Lexington, Mass., 1974).

tary aid to Angolan rebels were victories not only for the foreign policy preferences of liberals and the left but also for a Congress seeking to reassert its power over the imperial presidency. In contrast to Kennedy's short-lived opening to the left on nuclear weapons in 1963, the Democratic Congress in the mid-1970s was part of a genuine challenge to cold war assumptions about intervention and the use of conventional force. But this challenge was to fade toward the end of the decade.

The rubric New Politics did not, of course, imply wholly new issues or political constituencies. Peace, women's rights, environmentalism, and consumerism had all made earlier appearances as movements supported by middle-class to upper-middle-class constituencies, including students, women, and professionals. Yet the movements from the 1960s onward were unique in the postwar era and arose in rapid succession within a decade, empowered largely by the growth of the core constituencies, especially students and professionals. Perhaps more to the point is the contrast between the "old politics" of the New Deal Democratic party of urban machine politicians, union leaders, corporate managers, and southern power brokers and the new politics of decentralization, participation, and power for groups with issues hitherto not part of the traditional party structure and agenda.[15] As the New Deal coalition and agenda slowly disintegrated, helped along by New Politics activism, the Democratic party came to depend on the New Politics coalition for much of its popular support, especially during the Republican's eight-year occupancy of the executive branch from 1969 to 1977.

1975–1980: From Antimilitarism to the Nuclear Peril

Ironically, in the aftermath of the Vietnam War, peace groups and their cause nearly foundered on the political shoals of détente, despite the overall strength of New Politics issues. Nixon withdrew the troops as he produced SALT I. Nixon, Ford, and Congress kept defense spending in a pattern of relative decline from the war years. With the war over, the public seemed less concerned with national security and foreign policy. Broadly based appeals for a less aggressive foreign policy and smaller military budgets drew little support. Opponents of U.S. military and foreign policy were stymied.

---

15. Burkhart and Kendrick, in *New Politics*, make this point.

Antiwar groups that formed or prospered during the Vietnam War began to join in the middle and late 1970s with organizations that supported the antiwar movement but focused primarily on nuclear weapons. Although the coalition was loosely knit, one umbrella group that brought together organizations of various stripes was the Coalition for a New Foreign and Military Policy, which developed from two antiwar organizations. In 1976, the Coalition on National Priorities and Military Policy, which was started in 1969 to advocate redirection of military funding toward domestic needs, combined forces with the Coalition for a New Foreign Policy, which had been the Coalition to Stop Funding the War. Most of the major peace groups that would mobilize around nuclear weapons gathered under this organizational umbrella, but in 1976 its priorities were still human rights, cutting military aid, and normalizing relations with Vietnam, an agenda that did not sell well with the American public. Many constituent groups of the coalition experienced hard times in membership and funding. For example, SANE stressed demilitarizing foreign policy, economic conversion, meeting human needs, and the Middle East in the 1976 presidential campaign. By 1977 it, like several other groups, had bottomed out; in fact, SANE nearly closed because of financial difficulties. Gradually, however, the target for remobilization came into focus; it was nuclear weapons and arms control. By the end of the decade most peace groups were devoting more resources to the nuclear threat and less to other aspects of American militarism.

Even as many national peace groups foundered, another movement was taking shape here and there, across the country. The anti–nuclear power movement, organized primarily around local alliances, was an amalgam of environmentalists and antiwar activists.[16] Mass demonstrations and arrests took place at nuclear power plants in several states in 1977 and 1978. The accident at Pennsylvania's Three Mile Island facility in 1979 turned the movement's cause into a national nightmare overnight.

Meanwhile some activists had been making the connection to nuclear weapons as well, with demonstrations at such places as the Rocky Flats Nuclear Weapons facility in Colorado and the DOE in Washington. In late 1977, activists from peace and anti–nuclear

---

16. On the anti–nuclear power movement and its links to the nascent nuclear peace movement see Ann Morrissett Davidon, "The U.S. Anti-Nuclear Movement," *Bulletin of the Atomic Scientists*, December 1979, pp. 45–48.

power organizations came together to unite these strands of protest. The result, in 1978, was Mobilization for Survival, which had as its paramount goals an end to the nuclear arms race and an end to nuclear power. Presaging the nuclear freeze proposal, Mobilization issued a call for an international moratorium on nuclear weapons and power. It organized the decade's most significant protest over nuclear weapons and their proliferation (and also over nuclear power). The protest, at the 1978 special United Nations session on disarmament, drew over twenty thousand participants. The melding of antiwar concerns with opposition to nuclear technologies resuscitated the peace movement.

Events that transformed the revival into a national campaign came in rapid succession during the second half of Carter's presidency. In 1976, Jimmy Carter's election seemed a partial expression of New Politics sentiment because of his stances on women's rights, human rights, arms control, and the military budget. Carter entered office under a pledge to trim defense spending, which he did in his first budget plans. He even dared to cancel the B-1 bomber program in 1977, after a campaign by peace groups spearheaded by the Coalition for a New Foreign and Military Policy.[17] In 1978, SANE began to stress nuclear disarmament and initiated campaigns against the MX missile and for the SALT II treaty.

But public and congressional opinion were already being turned from the post-Vietnam caution and détente-inspired hopes toward the newly publicized Soviet nuclear buildup and Third World adventurism. Congress started to increase Carter's defense budgets, and the president began to toughen his stance on defense. Events and his own actions, however, would stick Carter between the rock of a revived cold war and the hard place of a nascent peace movement. The very events and policies that mobilized conservatives in favor of a military buildup also mobilized liberal activists. In 1979, Carter decided to proceed with full-scale development of the MX and deployment in the mobile "racetrack" range in Utah and Nevada. The geographic and public visibility of the MX plan was, along with the death of SALT II, the best ammunition the peace movement could expect.

17. The cancellation was the most concrete achievement of the post-Vietnam antimilitarism coalition up to this time; defeating such an important system instilled confidence for future initiatives. The campaign was focused less on the dangers of the B-1 as a nuclear weapons platform, though, than on the bomber as an outrageous waste of money that could be diverted to domestic needs.

In December of that year, the president led NATO into the theater nuclear forces decision to deploy Pershing IIs and cruise missiles in Western Europe. This move precipitated the resurgence of the European peace movement, which would begin to flood the streets of several NATO capitals in 1981, well before similar demonstrations could be mobilized in America.[18] In 1980 the Congress passed funding for development of the B-1 or another new bomber. The administration played its final nuclear card in the summer of that year. Having exhausted all budgetary and hardware options in this public relations campaign, Carter resorted to doctrinal warfare. In late August, Defense Secretary Harold Brown gave a speech before the Naval War College in which he revealed the administration's "new" nuclear strategy, embodied in the supposedly secret PD-59. The doctrine opted for limited nuclear options, or counterforce strikes, against Soviet targets. This strategy, which was intended to show that the Democrats were responding to the nuclear "window of vulnerability," only showed Republicans that their political message had struck a sympathetic public chord. Many liberal Democrats and peace activists discovered that the president and his party would not champion their cause in the new decade.[19]

Although macroeconomics apparently had more to do with Reagan's election in 1980 than did the politics of national security, he entered office with strong support, at both the elite and mass levels, for significantly greater military spending in general and for a strategic revitalization program in particular.[20] Reagan's election exacerbated the fears of liberals and of others who were for the first time

---

18. At about the same time in 1980 that Randall Forsberg was just formulating the freeze proposal, E. P. Thompson issued the Call for European Nuclear Disarmament, which became the credo of the nascent movement. As early as June 1981, about 100,000 marched in Hamburg. During two weekends in October, more than 850,000 protested in London, Bonn, Rome, Brussels, and Amsterdam, the capitals of the five NATO nations scheduled to receive cruise and, in the case of West Germany, Pershing missiles.

19. One example of an elite activist whose high hopes for Carter were dashed by late 1978 is Harold Willens, a Los Angeles industrialist. Disillusioned by Carter's turn to the right on arms control, especially his failure to attend the first United Nations special session on disarmament, to which Willens was a citizen-delegate, Willens and his California colleagues realized they could no longer expect initiatives from Carter or Congress. They decided to initiate grass-roots efforts, and Willens went on to lead the California freeze referendum drive. See *National Journal*, 18 September 1982, p. 1603.

20. In the 1980 Center for Political Studies National Election Study, economic responses dominated answers to the most-important-problem question. "Iran" was the only significant foreign policy response, as discussed on page 29.

alerted to the awesome potential of the nuclear arsenals. The combination of doctrines for fighting nuclear war, harsh cold war rhetoric, new strategic weapons programs, and rapidly increasing military budgets provoked a groundswell of activity by activists, old and new. The dramatic change in the cold war attested to by the demise of SALT II, the approval of the MX, and Reagan's election aroused both media attention and public concern. The transition from Carter to Reagan left a political legacy of a public rapidly polarizing over the issues of nuclear weapons, nuclear deterrence, and arms control. This division pitted the relatively united Republicans against New Politics liberals, with the New Deal leadership of the Democratic party foundering somewhere between. Liberals, movement activists, and the Democratic party needed new issues to revitalize their agenda and rebuild political coalitions: the emerging peace movement would soon become the catalyst for a resurgence of liberalism.

The Ideology and Program of the Emergent Peace Movement

Unlike the cold war coalition, the peace movement had no single locus of leadership and decision making. Consequently, its ideology was less neatly formulated and its agenda not as readily capsulized. Nevertheless, its various elements converged around a few dominant themes and arguments and developed a broad but coherent critique of nuclear warfare, weapons, and strategy.[21] If the two fundamental ideas behind cold war militarism have been peace through strength and containment, the analogous ideas for the peace movement have been negotiation and nonintervention. Peace through strength and containment entail a militarization of foreign policy; negotiation and nonintervention, the opposite. The peace movement consistently invoked the need for negotiation as the means to mitigate superpower conflict; hence its overriding emphasis on bilateral negotiations for arms control. But negotiations were also advo-

---

21. The claims made in this section are based on the author's own work and participation in the peace movement from 1979 to 1986 and on his extensive collection of mailings and literature from a wide variety of organizations, including Council for a Livable World, SANE, Physicians for Social Responsibility, Union of Concerned Scientists, Coalition for a New Foreign and Military Policy, Mobilization for Survival, Lawyers Alliance for Nuclear Arms Control, American Friends Service Committee, Center for Defense Information, Women's International League for Peace and Freedom, Greenpeace, and Common Cause.

cated for regional problems that supposedly fuel the arms race—those in Afghanistan, Central America, and the Middle East. Thus, too, where an option was the use of conventional force, the peace movement was decidedly against military intervention. Although many nuclear arms control and disarmament groups at the heart of the movement eschewed taking direct positions on issues outside of the nuclear dilemma, their reasons were primarily strategic and reflected no internal debate over U.S. policy—for example, toward Nicaragua.[22] Many other peace groups willingly and openly took positions on potential and actual cases of U.S. intervention and drew explicit connections between intervention and the possibility of escalation to nuclear conflict.[23]

Negotiation and nonintervention were the backdrop for more specific arguments against the nuclear arms buildup. First, the peace movement had as its de facto motto that there can be no winners in a nuclear war. This assertion combined two powerful arguments, one about morality and the other about utility. Nuclear war would be so indiscriminate in its destruction that it could not be morally justified. Furthermore, the probability of unlimited devastation meant that nuclear war could not serve any political purpose. Nuclear war would not be the continuation of politics by other means; it would be the end of politics and most everything else. The emphasis in the movement and the media on the actual and potential effects of nuclear weapons and war gave this argument weight and credence with the public.

Implicit in the no-winners argument was another premise of the movement: that the arsenals of the superpowers had long outgrown any reasonable size. Both sides were adding to what had for many years amounted to irrational overkill. Statistics about how many times over the superpowers could destroy the world were commonly cited. Overkill was not the movement's sole concern about the growing arsenals, however. Not only were there too many nuclear weapons, but the new ones being built were increasingly dan-

---

22. For example, Council for a Livable World does not take stands on issues other than weapons of mass destruction, but its mailings soliciting money for candidates have often contrasted candidates' positions on issues such as aid to the Contras and abortion because of the general liberal proclivities of its membership.

23. Mobilization for Survival, American Friends Service Committee, and Coalition for a New Foreign and Military Policy, for example; see *Nuclear Times*, October 1982, p. 6, for a roundtable of peace movement leadership opinion on the political wisdom of stressing the linkage between nuclear and other forms of militarism.

gerous. The upsurge in public concern and movement activity on matters of nuclear weapons and war was in part spurred by a new generation of weapons and a supposedly new strategy, exemplified in the MX missile and Carter's PD-59 doctrine. The MX represented the advent of purely counterforce weaponry, while PD-59 candidly acknowledged plans to fight a nuclear war. Together, the capability and the plan evinced a calculated willingness on the part of political and military leadership to use the weapons. The exposure of Reagan's version of nuclear strategy only exacerbated these concerns.

Furthermore, the weapons were becoming more expensive. If overkill existed and counterforce was counterproductive, then much of the money spent on the nuclear arsenal was unnecessary and could be better spent elsewhere in the national security budget or, better yet, in domestic programs. The peace movement frequently relied on graphic displays of the costs of nuclear programs and was especially fond of contrasting those costs with relatively puny social programs.

In essence these points amounted to an argument for minimum nuclear deterrence. The use of nuclear weapons can serve no useful purpose. They are provocative and expensive. Therefore the only tenable use is as a deterrent, and that mission can be accomplished with greatly reduced arsenals. Consequently, the superpowers had a compelling mutual interest in stopping and reversing the arms race. Ultimately, to secure the world from accident and miscalculation, nuclear weapons must be eliminated completely.

Conspicuously absent by contrast was an analogous discussion of the requirements of conventional deterrence and defense. If the nuclear peace movement was quietly anti-interventionist, it was also relatively mute on what was to be done about Reagan's conventional buildup and new conventional strategies. While the peace groups generally advocated a reduced defense budget, they emphasized cutting nuclear programs. Some within the movement were willing to talk about the need to refurbish conventional forces instead of the nuclear arsenal, but the movement was far from united on this matter. Some of the groups with stronger roots in the Vietnam War campaign often talked about cutting high-profile conventional programs such as aircraft carriers and jet fighters. By and large, however, the public relations effort of the mainstream movement kept its distance from any detailed critique of conventional forces, strategy, and requirements; for such a critique was unnecessary, at least as long as all attention remained focused on the nuclear nightmare. But the peace movement did not rely only on fear of Armageddon and opposition to an expensive buildup; it developed its own alternative agenda for national security.

Preventing production and deployment of new nuclear weapons systems was high on the list of movement priorities. This followed from the arguments about overkill and the dangers of counterforce: new systems were militarily unnecessary and would be far more destabilizing than their predecessors. New systems were also easy targets for media and public attention. And the list was a long one, including the MX, the B-1 bomber, the Trident II missile, the cruise missile (ALCM and SLCM), and the Euro-missiles (GLCM and Pershing II). In particular (as we shall see in Chapter 5), the MX came to symbolize the Dr. Strangelove logic and Rube Goldberg machinations of nuclear war planners. The peace movement also worked, when necessary, to halt the production of chemical weapons, such as binary gas artillery shells. Although chemical weapons received little attention from the media, when the votes for funding other weapons of indiscriminate destruction came up in Congress, the movement lobbied diligently.

While attempts to cut new weapons were intended as a technological restraint on the implementation of a first-strike strategy, the no-first-use proposal was meant to add an explicit moral and doctrinal restraint. McGeorge Bundy, George Kennan, Robert McNamara, and Gerard Smith, who came to be known as the Gang of Four, quickly responded to Euro-missile protest movements and rapidly shifting sentiments at home with a proposal that the NATO alliance declare a policy of no-first-use of nuclear weapons.[24] Ironically, many of the more conservative allies of the movement considered no-first-use to be perhaps the most radical item on the movement's agenda and opposed it, while other movement activists viewed it as an empty gesture that might detract from the freeze campaign and invite conventional rearmament.

Bilateral arms control, however, was the sine qua non of the peace

---

24. McGeorge Bundy, et al., "Nuclear Weapons and the Atlantic Alliance," *Foreign Affairs* 60 (Spring 1982): 753–68. The authors readily admit that one purpose of a no-first-use declaration would be "to meet the understandable anxieties that underlie much of the new interest in nuclear disarmament, both in Europe and in our own country" (p. 764). Even though no-first-use would be simply a declaratory policy, preventing no weapon from being built and deployed, the Gang of Four repeatedly emphasizes that any such policy would necessarily go hand in hand with conventional rearmament: "An allied posture of no-first-use would have one special effect that can be set forth in advance: it would draw new attention to the importance of maintaining and improving the specifically American conventional forces in Europe" (pp. 760–61).

movement, and the comprehensive nuclear weapons freeze proposal was the centerpiece of the movement's programmatic and public relations effort. The nuclear freeze proposal was a straightforward call to halt the nuclear arms race through a bilateral cessation of the testing, production, and deployment of all nuclear weapons and nuclear weapons delivery systems. The freeze became the positive policy embodiment of the movement's beliefs. Peace groups had always been accused of wanting only to slash at the American military budget; now, in the freeze proposal, they had a simple idea with the same direct appeal as peace through strength. The freeze mobilized public opinion, and the widespread reporting of public sentiment was the movement's most effective weapon. Of course there were other arms control measures that ranked high on the movement's agenda. Second to the freeze came completion of the comprehensive test ban treaty, which the superpowers had started to negotiate, but which the Reagan administration decided not to pursue. In addition, the movement sought to protect existing treaties, particularly SALT II and the ABM treaty. But it was the freeze proposal that coalesced the movement, "democratized" the arcane world of national security and nuclear strategy, and offered a simple and direct solution to the "nuclear nightmare."

The Mobilization of the Freeze Movement

Various unilateral and multilateral moratoriums on the production and deployment of nuclear weapons have been proposed over the past twenty-five years. A revival of the moratorium idea began in 1979 during the SALT II debate. In the fall, during the Senate Foreign Relations Committee's consideration of the treaty, Senator Mark Hatfield offered an amendment that would have had the United States and the Soviet Union negotiate a moratorium on the testing, development, and deployment of strategic nuclear weapons. The amendment was defeated; the treaty never made it to the Senate floor. As Hatfield later noted of his amendment, "I do not think it caused a blip on the political screen."[25] About the same time Hatfield was tilting against windmills in the Senate, a defense analyst

---

25. Hatfield discusses his amendment and his subsequent efforts for the freeze in his congressional testimony on arms control; see U.S. Senate, Committee on Foreign Relations, *Nuclear Arms Reduction Proposals*, 97th Cong., 2d sess., 29 and 30 April, 11–13 May 1982, p. 147.

and peace activist named Randall Forsberg was formulating a bilateral freeze proposal as a way to unite the nascent peace movement. Forsberg initially conceived of dual goals to be pursued simultaneously: a bilateral nuclear weapons freeze and a bilateral U.S.-USSR agreement not to intervene in developing nations. Because two goals would complicate the issue and because the nuclear proposal seemed more marketable, Forsberg dropped the nonintervention idea.[26] In January 1980, Forsberg introduced the proposal to a meeting of about thirty peace groups, which was sponsored by the Fellowship of Reconciliation. Several weeks later, several peace groups, including the American Friends Service Committee, the Fellowship of Reconciliation, Clergy and Laity Concerned, and Forsberg's newly formed Institute for Defense and Disarmament Studies published the proposal as the *Call to Halt the Nuclear Arms Race*. The call advocated a verifiable and bilateral "freeze on the testing, production, and deployment of nuclear weapons and of missiles and new aircraft designed primarily to deliver nuclear weapons."[27] The appeal was constructed to elicit support from the broadest possible audience. Two characteristics made the freeze potentially salable to the concerned public and relatively unassailable by conservatives: first, unlike many arms control proposals, it was simple and direct; second, it was explicitly bilateral.[28]

Although a freeze plank was quickly rejected by a hawkish Democratic Convention in August of 1980, over a dozen prominent peace and liberal defense groups met in September and agreed to coordinate their efforts around the freeze.[29] Early success took place at the local level through endorsements by Vermont town meetings, which the American Friends Service Committee helped organize. Randall Kehler, a Vietnam-era activist and draft resister, led a nuclear freeze referendum campaign in western Massachusetts state senatorial dis-

26. See Forsberg's forward to the Melinda Fine and Peter M. Steven, eds., *American Peace Directory* (Brookline, Mass., 1984). David S. Meyer, *A Winter of Discontent* (New York, 1990) provides a far more detailed history of the origins of the freeze proposal and its acceptance by the peace movement.
27. American Friends Service Committee, *Call to Halt the Nuclear Arms Race* (Philadelphia, n.d.).
28. Public appeal is discussed in the call, which states that in contrast to a bilateral freeze, "campaigns to stop individual weapon systems are sometimes treated as unilateral disarmament . . . [and] the pros and cons of SALT II are too technical for the patience of the average person."
29. This meeting included more mainstream organizations such as the Physicians for Social Responsibility, the Federation of American Scientists, and the Center for Defense Information.

tricts during the 1980 elections, garnering majority support for the freeze proposal in thirty of thirty-three towns that voted for Reagan. Soon after Reagan's inauguration, peace activists held the first freeze conference in March 1981. This meeting established the freeze as a grass-roots campaign and set up an educational clearinghouse for freeze information, to be run by Kehler in St. Louis; it also united various groups around the freeze as the central programmatic effort for the whole nuclear weapons peace movement.[30] With the adoption of the freeze by both peace groups and more moderate arms control organizations, and the initiation of an independent freeze campaign, the movement had found its unifying principle and its vehicle for popular mobilization.

During its first convention, the freeze campaign gave highest priority to the education and organization of the public. Washington would have to wait until the public was mobilized, from the grass-roots upward. The battle for public opinion was driven by an educational campaign that sought to introduce the issue into every possible forum across the country, including schools and churches. Leading the way for the mainstream religious community was the National Council of Churches with its 1980 endorsement of the freeze. On Veterans' Day 1981, teach-ins on over one hundred college campuses, organized by the Union of Concerned Scientists, attracted media coverage and spurred organizational efforts by college students. The Physicians for Social Responsibility, revived in the late 1970s, organized a series of public symposia on the medical consequences of nuclear weapons and war, which they called the "final epidemic."

Organizational Expansion

The growth and multiplication of public education events and programs went hand in hand with an explosion in the number and kind of organizations involved in the new peace movement. Existing groups grew dramatically and new organizations appeared everywhere. The core of the existing organizational base was composed of groups with religious affiliations, such as the American Friends

---

30. The decision to focus on the freeze was controversial. Some groups, such as Mobilization for Survival, while endorsing the freeze, refused to put aside related issues, such as intervention. Frances B. McCrea and Gerald E. Markle, *Minutes to Midnight* (Newbury Park, Calif., 1989), pp. 103–4.

Service Committee; pacifist groups such as the Fellowship of Reconciliation; groups dominated by scientists, such as the Union of Concerned Scientists; and liberal defense policy/arms control public interest groups like SANE and Council for a Livable World. All these groups grew and prospered. The direct-mail and lobbying groups also swelled with new members and generous donation levels. The Council for a Livable World, a PAC for the Senate, grew from 12,000 members in 1980 to over 60,000 in 1984. Its success allowed it to form an educational fund and then a PAC for the House in 1982. Its donations to candidates approximately doubled every two years, from about $250,000 in 1980 to just about $500,000 in 1982 to over $1 million in 1984, making it one of the largest public interest PACs. SANE's membership grew from about 20,000 in 1980 to about 80,000 by 1984. As Michael Mawby of SANE put it, "Reagan's been great for us even if he's bad for the world."[31]

Not only did existing organizations prosper, new groups formed rapidly. From 1979 to 1984 the number of arms control and peace groups operating at a national level nearly doubled.[32] Proliferation at the local level was greater still. And many of these new organizations were unlike the established ones in several ways. First, new peace groups tended to be concerned predominantly or solely with nuclear weapons and arms control, whereas older groups often had broader agendas. Second, the majority of new groups were action oriented rather than educational. Not only did new peace groups form; many New Politics causes latched onto the skyrocketing peace movement and its issues. Environmental groups such as the Sierra Club and Friends of the Earth began to slant more fund-raising appeals toward nuclear war issues. Common Cause, not usually associated with national security policy, started campaigns against nuclear weapons programs. Women's organizations such as NOW and the League of Women Voters endorsed the peace movement's efforts.

The proliferation of new groups and allies was also indicative of

---

31. Quoted in Steven Pressman, "Nuclear Freeze Groups Focus on Candidates," *Congressional Quarterly Weekly Report*, 5 May 1984, pp. 1021–24.
32. Proliferation data are based on counts made from two sources: Institute for Defense and Disarmament Studies, *Peace Resource Book* (Cambridge, Mass., 1986); and Forum Institute *1983 Handbook* (Washington, D.C., 1983). A count based on similar sources also revealed that only 16 percent had formed between 1975 and 1979, see McCrea and Markle, *Minutes to Midnight*, p. 119.

the peace movement's strongest constituency: liberal members of the New Class. Groups with explicit ties to professional and technical vocations dominated, numerically, the list of new organizations. An incomplete list illustrates the nature of their constituencies: Physicians for Social Responsibility (revived in 1979), Educators for Social Responsibility (1981), Lawyers Alliance for Nuclear Arms Control (1981), Psychologists for Social Responsibility (1982), Professionals' Coalition for Nuclear Arms Control (1984), Computer Professionals for Social Responsibility (1983), Communicators for Nuclear Disarmament (1981), Business Executives for National Security (1982), Musicians Against Nuclear Arms (1982), High Technology Professionals for Peace (1981), Social Workers for Peace and Nuclear Disarmament (1981), Performing Artists for Nuclear Disarmament (1982), Architects for Social Responsibility (1982), Lawyers Committee for Nuclear Policy (1981), United Campuses to Prevent Nuclear War (1982). Some groups represented more than the collective expression of concerned individuals within a profession; for example, the Social Workers for Peace and Nuclear Disarmament was an official committee of the National Association of Social Workers. The National Education Association endorsed the freeze, along with the American Federation of State, County, and Municipal Employees (AFSCME) and the U.S. Conference of Mayors. Many of these professions and groups have strong ties to the public sector and were hurt by Reagan's efforts to cut federal support for social programs.

Monetary support for the peace movement was forthcoming, however, from still another type of organization that involved itself heavily in the new movement: educational and philanthropic foundations. Money began to pour forth from such establishment institutions as the Carnegie and Ford foundations and the Rockefeller Fund as well as from more liberal institutions such as the MacArthur Foundation. Moreover, new groups were started with the sole purpose of funding projects related to international security. Between 1982 and 1984, foundation funding for peace, arms control, and international security increased 200 percent, from $16.5 million to $52 million.[33] The bulk of this money funded academic programs and research, but much also went toward more immediate and broader public education programs and mass media productions.

33. McCrea and Markle, *Minutes to Midnight*, p. 117.

The National Mass Media Teach-in on the Nuclear Nightmare

The news and entertainment media reacted to, and in turn amplified, movement activities and public concern. The mass media can create and spread the perception of movement strength or weakness. In this case, media activity was crucial in augmenting the power of the nuclear weapons issue and the movement behind it.

The unprecedented nature of the Reagan administration's rearmament program and rhetoric on arms control and nuclear warfare, coupled with the obviously high level of public concern, triggered a wave of journalistic coverage and dramatic productions. Perhaps the earliest entry in this field was the five-part CBS Reports series entitled "The Defense of the United States," broadcast during the summer of 1981. It covered most aspects of American defense, both conventional and nuclear. The highlight of the series was the episode that simulated the effects of a nuclear attack on Omaha, Nebraska, the home of the Strategic Air Command Headquarters. The next summer, NBC followed with its white paper, "Facing Up to the Bomb."

In 1982, Jonathan Schell's *Fate of the Earth* quickly became a bestseller. In it he uses a poignant discussion of the probable effects of nuclear weapons to drive home the moral necessity for action to end the threat. The book had been published previously as installments in the *New Yorker*. Another pillar of the eastern media, the *Boston Globe*, produced a special Sunday section in October of 1982 devoted entirely to the nuclear arms race. Titled "War and Peace in the Nuclear Age," it featured articles on the history of the arms race, the effects of nuclear weapons, and arms control negotiations. It later won a Pulitzer prize.

Finally, and certainly most controversial, there was ABC's "The Day After," aired in November 1983. This drama featured Jason Robards and a cast of hundreds coping with the terrifying consequences of a nuclear war, this time in Kansas. Well before the broadcast, conservatives charged that it would pander to fears that would engender support for disarmament. ABC sanitized the broadcast by following it with a debriefing featuring Ted Koppel and a panel of distinguished experts of various persuasions.

Whether through print or picture, documentary or drama, during these years the U.S. public was exposed to information and images about the reality and potential of nuclear weapons, in an unprecedented blitz of media coverage. Perhaps the most significant impact came from the ubiquitous analyses and depictions of the effects of nuclear weapons. For the first time, the majority of Americans were confronted with the potential of nuclear conflict, unvarnished by the bland reassurances of a government publication or film on fallout shelters.

Public Opinion and Political Victories

Movement organizations and the mass media tapped into and created a nationwide surge in public opinion. Although Reagan entered office with manifest popular support for an increase in defense spending and a nuclear weapons rearmament program, public opinion, led by the movement's efforts, very quickly swung back toward more dovish positions.[34] Early indications showed great support for moratoriums on arms production and cuts in the nuclear arsenals. A May 1981 Gallup poll had 72 percent of Americans in favor of a superpower agreement not to build any more nuclear weapons.[35] In a November Gallup poll, 76 percent of Americans supported George Kennan's proposal for a bilateral 50 percent across-the-board reduction in nuclear weapons.[36] Nuclear arms control appeared to be the weak link in Reagan's cold war politics—and perhaps in the Reagan revolution as a whole. While most of the public favored getting tough with the Soviets and spending more on defense, an overlapping and substantial majority also feared the consequences of a nuclear arms race and wanted arms control agreements. The freeze proposal breached this gap in the administration's national security agenda, which spurned arms control efforts.

Once the movement put its full weight behind the freeze, the freeze attracted widespread support and became the barometer of the movement's public appeal. As early as May 1982, national polls had 70 percent or more of Americans backing a freeze.[37] Support was

---

34. A comparison of two analyses of American public opinion and national security coauthored by pollster Daniel Yankelovich yields findings complementary to those in this section: Daniel Yankelovich and Larry Kaagan, "Assertive America," *Foreign Affairs, America and the World, 1980* 59 (1981): 696–713; and Daniel Yankelovich and John Doble, "The Public Mood," *Foreign Affairs* 63 (Fall 1984): 33–46.
35. *Gallup Reports* 188 (May 1981).
36. Ibid. 196 (January 1982).
37. Judith Miller, "72% in Poll Back Nuclear Halt If Soviet Union Doesn't Gain,"

remarkably consistent across class and partisan divisions. In Gallup surveys from November of 1982, 71 percent favored a bilateral freeze, while 45 percent (versus 55 percent) even supported a freeze "whether or not the Soviet Union agrees to do the same."[38]

Finally, the support for more military spending also faded. Whereas in 1981 more than 51 percent thought we were spending too little, while 15 percent said too much, by November 1982, 16 percent said too little and 41 percent said too much.[39] The freeze and arms control attracted much greater overall support and more nearly equal support across political divisions. Typical of poll results were the following: whereas 19 percent of Republicans, 40 percent of Democrats, and 37 percent of Independents were for a cut in defense spending, 55 percent of Republicans, 60 percent of Democrats, and 60 percent of Independents favored an immediate freeze.[40]

The political impact of the peace movement was evident in more than public opinion polls. The movement and the nuclear freeze campaign also introduced the freeze proposal for consideration by local, city, and state governments and won in the vast majority of cases. By mid-1982, over four hundred town meetings, over two hundred city councils, and nine state legislatures had voted in favor of the freeze. But perhaps the most dramatic indicator of the movement's power came in June 1982 at the New York City nuclear freeze rally, which drew a crowd estimated at seven hundred thousand. It was, by some measures, the largest single political rally in the country's history.[41] The New York City rally was a politically powerful manifestation of the explosive transformation of public concern and organizational efforts into a nationwide movement.

This success with public opinion and local and state legislation did not go unnoticed by national politicians in both parties. Moreover, the leadership of the freeze campaign was being drawn toward a frontal assault on Washington. The freeze was, after all, not a local, but a national, issue. The showdown between the peace movement

---

New York Times, 30 May 1982, pp. 1, 20.
  38. *Gallup Reports* 206 (November 1982); and ibid. 208 (January 1983).
  39. Ibid. 208 (January 1983).
  40. Figures obtained from the following computer-based data set: Chicago Council on Foreign Relations, *American Public Opinion and U.S. Foreign Policy, 1982* (Ann Arbor, Mich., 1983).
  41. Perhaps the rival for the claim would be the 24 April 1971 antiwar rallies, which drew as many as seven hundred thousand to Washington and three hundred thousand to San Francisco.

and legislators in the capital would happen sooner rather than later. In particular, leaders in the Democratic party were looking to nuclear weapons and arms control for a pivotal issue, a weak link in the Reagan agenda. Circumstances and the efforts of the peace movement showed within a year after Reagan's inauguration that the issue of nuclear weapons and strategy held perhaps the greatest potential for a first attempt to reverse or mitigate the political consequences of the 1980 elections. As the most salient issue of the early 1980s, the nuclear dilemma became the linchpin for the revitalization of a liberal coalition led by the New Politics groups. In this way, the peace movement, with the freeze as its main weapon, would lead the counterattack against not only Reagan's military policies but the overall Reagan revolution as well. The elections of 1982 would test the power of the movement and the freeze initiative and, more important, forge strong links between the peace movement's issues and the electoral strategies of many Democratic politicians, as we shall see in Chapter 5.

4

The Origins of Military Reform

Amid the tumult of the fight between the freeze movement and the Reagan administration, a third position on national defense policy began to emerge. News reports featured stories on weapons that cost too much and did not work. Books and articles appeared every month bemoaning the lackluster performance of the American armed services in war and peace. Representatives and senators regularly proclaimed the desperate need for significant changes in the way the Pentagon does its business. Generally avoiding the controversy surrounding the nuclear arms race, a coalition of members of Congress, public interest groups, military officers, and civilian analysts began to champion reform of conventional war strategy, conventional weaponry, military organization, and the arms procurement process. Although it would take longer for military reform to become a force to be reckoned with in American politics, with Congress as its institutional base, it was eventually to eclipse both the buildup and the peace movement as the dominant defense policy agenda.

The defense reform movement of the 1980s is not, of course, without antecedents scattered over the course of American history. Although reform efforts have been more numerous since the advent of the postwar military establishment, defense reform dates back to the first years of the republic, when the Congress of the Confederation ordered an investigation of waste and fraud in the use of public property during the revolutionary war—the first look at procurement reform.[1] At the beginning of the twentieth century, disorganization and interservice rivalry during the Spanish-American War inspired the Root Reforms, which included the establishment of new chains of command and interservice coordination—an early reorganizational effort. The Nye Special Committee investigated the relations between the military and the munitions industries in the 1930s.

World War II, however, placed unprecedented demands on American military capabilities and thus led to more frequent and sweeping reform efforts. The National Security Act of 1947 created the military and foreign intelligence apparatus in place today. It mandated the creation of the National Military Establishment headed by the secretary of defense, which established unified civilian command of the military. Other provisions gave statutory authority to the Joint Chiefs of Staff, a separate air force, and Munitions and Research and Development boards. The act also created the National Security Council and the CIA. Amendments and changes in 1949 and 1953 all sought to strengthen the authority of the defense secretary over the separate services.

The two most recent major efforts before 1980, one organizational and the other on procurement, prefigured contemporary reforms, in kind if not in scope. The first of these, the 1958 reorganization law already mentioned, was in part a product of a cold war mobilization by the Democrats and Eisenhower's attempt to respond to it. Its chief provisions gave greater authority to the defense secretary, especially over operational command, and augmented the advisory role of the Joint Chiefs of Staff. Again, the principal purpose was greater unification of command and control. This structure remained largely intact until the mid-1980s.

In 1970, growing opposition of congressional Democrats to the Vietnam War and the strain the war put on governmental resources led President Nixon to appoint a blue-ribbon panel, which came to be called the Fitzhugh Commission, to examine the procurement process.[2] (Its recommendations would be echoed by the blue-ribbon Packard Commission of the 1980s, discussed in Chapter 7.) After the

---

1. David C. Morrison mentions this precedent and discusses others in "The Defense Reform Merry-Go-Round," *National Journal*, 22 March 1986, pp. 718–19.

2. See Thomas McNaugher, "Weapons Procurement," *International Security* 12 (Fall 1987): 88–89, on the Fitzhugh Commission and also the Packard reforms.

Fitzhugh report, Deputy Secretary of Defense David Packard implemented a series of reforms that have been the basis of the procurement process ever since, at the heart of which is the Defense Systems Acquisition Review Council (renamed the Defense Acquisition Board in 1987). A weapon must pass muster at three stages before it goes into full-scale production, the emphasis being on operational testing, or "fly before you buy."

Since Packard's 1970s reforms, some thirty reports have recommended other improvements and changes in military procurement and organization. What these first two major reform efforts share with those of the 1980s is the emphasis on procurement and reorganization, rather than tactics and training. What they do not share is duration and scope of conflict. Whereas most past efforts were brief and limited in breadth and degree of conflict, the prolonged wave in the 1980s involved more institutions and produced a wider array of proposals and legislation.

The Intellectual Foundations of Military Reform in the 1980s

The 1980s reform effort actually originated with analysts inside the military in the late 1970s. By the early 1980s, the debate gradually entered the popular press, as military officials, civilian analysts, and journalists took up the cause. Beginning with Gary Hart, politicians began to sound the call for reform, and the movement in Congress grew steadily. Almost all the arguments proffered in the cause of military reform have roots in the work of Pentagon employees or former employees and stem from one of four central areas of concern: minds, machines, money, and management.

Perhaps the most fundamental motivation for military reform was the perceived need to examine and reformulate military doctrine so as to reassert the primacy of the military mind, with its capacity for innovation, over bureaucratic routines and the related tendency to "fight the last war." Reformulators put forward several overlapping critiques of traditional and postwar U.S. defense strategy which had in common a renewed emphasis on conventional warfare. Reformers claimed that emphasis on nuclear strategy and weaponry is misleading, if not dangerous. Prudence and history show that real national interests depend on the successful application of conventional power. Because Reagan's nuclear buildup created a polarized debate over the nuclear arms race, no doubt many reformers saw conventional defense as an alternative focus that would allow them to appear concerned about national security while avoiding some of the "theological" contentiousness of the nuclear debate. While reformers agreed on the priority of conventional strategy, there was some disagreement about the geopolitical mission of the reformed military. In a split between contemporary versions of "Europe-first" and "Asia-first," some saw Europe as the keystone in the battle of East versus West, while others argued that recent history shows that the periphery, and not Europe, is where U.S. force is used.[3]

The most widely debated reform in conventional strategy is "maneuver" warfare, which can best be defined by contrast with its opposite, "attrition" warfare. Beginning with the Civil War and continuing through Vietnam, the U.S. military establishment has relied on relative advantages in material and technological resources to overpower, wear down, and outlast the enemy. The failure of this approach in Vietnam and the perceived inability to compete quantitatively with the Eastern bloc led many strategists to consider new approaches to offense and defense in combat. Noteworthy among them was then–air force colonel John Boyd.

After serving as one of America's finest fighter pilots in the Korean War, Boyd returned to Nellis Air Force Base to teach air war tactics. His influence began to spread with his pathbreaking "Aerial Attack Study," which set a standard for air combat doctrine and is the basis for his subsequent work. In the 1960s, Boyd broadened his purview to all of warfare and prepared a five-hour briefing, which he delivered many times before audiences of military brass and analysts in the 1970s. With examples from ancient Greece to the twentieth century, his "Patterns of Conflict" study outlines the correlates of military victory. His fundamental conclusion is that success in combat has depended on flexibility, deception, and mobility. Victory belonged not to those who met the enemy head on in a slugging match but to those who confused and outwitted him by staying one mental and tactical leap ahead. Boyd draws upon the work of Clausewitz and adds a corollary. From his work on air combat, Boyd

---

3. Gingrich and Reed refer to this as the "blue water" versus "continental" debate; see Newt Gingrich and James W. Reed, "Guiding the Reform Impulse," in Asa A. Clark IV, *The Defense Reform Debate* (Baltimore, 1984), p. 41. See also Jeffrey Record, "Implications of a Global Strategy for U.S. Forces," in ibid., pp. 147–65; and Stansfield Turner, "Toward a New Defense Strategy," *New York Times Magazine*, 10 May 1981, pp. 15–17.

developed the concept of observation-orientation-decision-action (OODA) cycles that face all adversaries in battle, whether in air, on land, or at sea. Combat is composed of countless and competing OODA loops. By staying one or more steps ahead of an adversary's loop, one creates Clausewitzian "friction" for the opponent. This disruption of the enemy's plans and actions has led, historically, to his organizational and spiritual breakdown. Boyd then draws broader conclusions for contemporary U.S. defense, in particular that attrition should be replaced by tactics and equipment suitable for maneuver warfare.[4] Maneuver would entail lighter, more mobile forces guided by more decentralized command and used to debilitate rather than crush the enemy.

Work such as Boyd's found friends in the military reform movement. Academics and analysts began to push manuever strategy.[5] The army adopted manuever tactics in its field manual FM-100-5 in the early 1980s. Its most important application is in the AirLand battle plan for the defense of Europe, a program the army shares with the air force. From Boyd and his allies, military reformers got their specific emphasis on new conventional tactics that stress mobility, flexibility, and decentralization of command.

Although the debate about strategy certainly came first, a strategic corollary, force structure, achieved an independent and greater importance in military reform. The arguments about weaponry which followed from those about strategic doctrine quickly became issues in and of themselves, testifying to the need to eliminate waste and build better weapons, no matter what doctrine should prevail. The first arguments were general ones that followed directly from the emphasis on conventional arms and implied an obvious, if unspecified, reordering of priorities away from nuclear programs and toward conventional weapons and programs. The advocates of maneuver warfare of course had more specific weapons and programs in mind. Mobility would require the greater use of light infantry units, whereas the present army is composed primarily of heavy or armored divisions. Maneuver also depends on greater use of air

---

4. For more on Boyd's work see James Fallows, *National Defense* (New York, 1981), pp. 27–31; and Jacob Goodwin, *Brotherhood of Arms* (New York, 1985), pp. 206–8.

5. On maneuver versus attrition see Edward Luttwak, "The American Style of Warfare and the Military Balance," *Survival* 21 (March/April 1979): 57; Steven L. Canby, "U.S. Defense Policy," *AEI Foreign Policy and Defense Review* 1, no. 3 (1979): 23–36; William S. Lind, "The Case for Maneuver Doctrine," in Clark, *Defense Reform Debate*, pp. 88–100.

power for close air-to-ground support and, especially, for air transport. For example, Gary Hart called for smaller aircraft carriers armed with aircraft capable of vertical takeoff and landing. Most widely debated in public and in government, however, were the relationships and trade-offs among the quantity, quality, and effectiveness of weapons. The defense insider perhaps most closely associated with the reformist position in the quality and quantity debate is Pierre Sprey, a former Pentagon weapons planner and colleague of Boyd and a private defense consultant.

Recent decades have seen technology become a driving force in weapons production. Many weapons programs are technically superannuated before production ends. Follow-on programs incorporate an ever-increasing degree of high-tech equipment. Advocates of technological superiority argue that we must exploit our advantages to compete with the Soviets, whose advantages are cheap labor and mass production. The United States also faces budgetary constraints that compel reliance on technological superiority over quantitative competition. Defense Secretary Weinberger stated the case this way: "The United States continues to rely on its superior military technology to offset the numerically larger forces threatening its security interests. We and our allies have never advocated a conventional military buildup that matches the Soviet bloc's numbers soldier for soldier, tank for tank, or aircraft for aircraft. Instead we have depended on superior military technology, and on better readiness, training, leadership, and better educated people steeped in freedom with all of the inestimable advantages that brings, to compensate for quantitative disadvantages. Modern technology makes our systems more effective and more survivable."[6]

Reformers such as Sprey complain that any advance in, for example, electronics, avionics, or armor has been its own justification for inclusion in the next weapon, a process known as "gold-plating" and defined as the imperative and practice of putting the most advanced technology on a system whether or not it is practical or prudent. Sprey agrees with advocates of high-tech that "there is no inherent contradiction between quality and quantity in weapons. But our now-entrenched defense of high-cost, high-complexity programs blocks us from using advanced, brilliantly-simplifying tech-

---

6. U.S. Department of Defense, *Report of the Secretary of Defense Caspar W. Weinberger to the Congress on the FY 1986 Budget, FY 1987 Authorization Request, and FY 1986–1990 Defense Program*, 4 February 1985, p. 29.

nology to achieve the large increases in both quality and quantity of weapons that the nation needs more desperately every year."[7] Sprey blames much of the high-tech bias on the use of theoretical "paper" analyses of systems. He relies on combat and testing data to show that existing cheaper alternative systems are more effective and efficient than current top-of-the-line U.S. and Soviet weapons and platforms. The often-cited example is the F-15 fighter versus its less expensive and less complex competition, the F-16. One-on-one, the F-15 may be the most lethal fighter in the world, but its sophistication suffers from the three important consequences cited by reformers. First, the advanced electronics and weapons and the high-powered engines require much servicing; hence the plane has a very high service-to-flight-time ratio. Second, the expense of this sophistication means that relatively few can be purchased, so each plane, instead of fighting one-on-one, would be facing several MIGs. Third, some of its technology is not necessarily relevant to the F-15's role as a combat fighter: most dogfights do not take place at such high speeds, and its weight limits its range and maneuverability.[8] In contrast, the F-16 has a better efficiency record; costs about half as much, so more can be procured; and has weapons and performance characteristics better suited for an air-superiority fighter. The "fighter mafia," including John Boyd, Pierre Sprey, and Franklin Spinney, pioneered the critiques of high-tech fighters in their push for the development of what would become the F-16.[9] This same kind of critique was then applied to other systems, such as the M-1 tank and the Bradley personnel carrier. In general, these reformers advocated greater use of battlefield-tested and relevant technologies so as to stretch the defense budget to provide larger numbers of more reliable weapons. This posture gave them the appearance of actually wanting more and better systems while simultaneously trimming military spending and the deficit. The gold-plating cri-

---

7. Pierre Sprey, "The Case for Better and Cheaper Weapons," in Clark, *Defense Reform Debate*, p. 205.
8. On both gold-plating and the contrasts between the F-15 and the F-16 see Fallows, *National Defense*, chap 3; and Goodwin, *Brotherhood of Arms*, chap 8.
9. Sprey, as an analyst for the Pentagon, allied himself with Boyd and his work. As a result, he produced the concept paper for a jet fighter that incorporated Boydian principles of maneuver, tested technologies, and freedom from gold-plating for what would eventually become the F-16. See Fallows *National Defense*, pp. 95–106. For work by Sprey see Pierre Sprey and Jack Merrit, "Quality, Quantity or Training," *U.S.A.F. Fighter Weapons Review* 27 (Summer 1979): 9; and Sprey, "Mach 2," *International Defense Review*, 8 (1980): 1209–12.

tique, best exemplified by Sprey's work, spurred congressional efforts to overhaul Pentagon procurement practices.

But the fixation on technological supremacy has insidious effects beyond quality/quantity problems. A Pentagon systems analyst named Franklin Spinney demonstrated that longstanding Pentagon procurement practices systematically undermine the nation's ability to translate long-term plans into actual military forces. Spinney came to national attention when he made the cover of *Time* in March of 1983. His briefings of congressional committees, despite Pentagon resistance, provided early impetus to reform.[10] Taken together, Spinney's two Pentagon briefings, "Defense Facts of Life" (1980) and "The Plans/Reality Mismatch" (1982), describe a pattern endemic in military planning and budgeting,[11] which leaves the military, despite larger budgets, with a less-prepared, smaller, and more-expensive arsenal. Three assumptions of planning set the cycle in motion. First, Pentagon planners assume that maintenance and operations costs will decline as a percentage of overall expenditures. Second, they plan under the assumption that the military budget will grow steadily for several years. Third, they repeatedly assume that procurement costs decline as the acquisition process proceeds. None of these assumptions is borne out. In fact, the opposite usually obtains.

These assumptions are particularly pernicious given the Pentagon's infatuation with high-tech weapons platforms. Short-term decisions to modernize with these expensive systems squeeze the operations and maintenance budget because there simply is not enough money for both procurement and increased operations and maintenance, not because these new capital-intensive systems cost less to operate and maintain. These systems then fail to decline in per-unit cost as the production run reaches maturity; usually costs grow, or else decline at a much slower rate than anticipated. These weapons and platforms end up requiring more maintenance, which absorbs more money. Finally, as they always do, budget cuts come

---

10. "The Winds of Reform," *Time*, 7 March 1983, pp. 12–30. His internal Pentagon briefing from 1980 ("Defense Facts of Life," presented to the Subcommittee on Manpower and Personnel, May 1980) was delivered before a Senate Armed Services subcommittee only after Senator Sam Nunn got it declassified, much to Weinberger's dismay. Senator Grassley, despite resistance from Senator Tower and refusals by Weinberger, eventually got Spinney before the Armed Services committee again to deliver his second internal briefing in 1983 ("Defense Plans/Reality Mismatch").

11. These two briefings, previously available separately as congressional testimonies, have been combined in a book: Franklin C. Spinney, *Defense Facts of Life: The Plans/Reality Mismatch*, ed. James Clay Thompson (Boulder, Colo., 1985).

and upset the original long-term planning. Instead of canceling the program, the production run is stretched out, raising costs even more. The whole process systematically drains resources from maintenance and training, or readiness. The result is a smaller, undermaintained, possibly undermanned, and more expensive arsenal undergoing modernization at a glacial pace. Furthermore, Spinney shows that none of the negative consequences flow directly from budgetary restrictions; budgetary restraint only aggravates a dynamic already in motion.

The Reagan administration and Pentagon officials vigorously denied Spinney's conclusions, arguing that Spinney's work was historical and therefore inapplicable to the Reagan era, during which efforts were underway to remedy such problems. Military reformers did not accept the administration's response to Spinney, and his work became a basis for congressional work on military budgeting and procurement reform. An interesting comment on the political fortunes of the reform movement is that Spinney's 1980 briefing received little attention, whereas his 1983 briefing, which delivered basically the same message, was the prop for a congressionally sponsored media showcase on the Pentagon's problems.

Other reform issues ostensibly have little to do with money and resources. Some reformers argue that greater benefits would accrue from nonmaterial changes in the military—changes in the organization and nature of military leadership. Virtually all these proposals seek to alter the bureaucratic nature of the military establishment, in particular, to limit interservice rivalry and reassert soldierly skills above managerial talents. Most reformers aimed their reorganizational efforts at the top of the military bureaucracy: at the Joint Chiefs of Staff and the overall command structure.[12] One important voice was former JCS chair General David C. Jones. Although his intellectual contribution was less original than those of Boyd, Sprey, and Spinney, Jones, along with Army Chief of Staff Edward C. Meyer, did initiate the debate, and his and Meyer's status lent it credibility.[13] The primary stated motive was to vitiate the effects of

---

12. For a history of postwar military organization and reform efforts see Victor H. Krulak, *Organization for National Security* (Washington, D.C., 1983). A good overview of the debate on the JCS and civil-military relations is MacKubin Thomas Owens, "American Strategic Culture and Civil-Military Relations," *Naval War College Review* 39 (March/April 1986): 43–59. For an argument against JCS reform see Owens's "The Hollow Promise of JCS Reform," *International Security* 10 (Winter 1985–86): 98–111.

13. David C. Jones, "What's Wrong with Our Defense Establishment?" *New York*

interservice rivalry on planning and command. Jones based his critique on a broad appraisal of mostly postwar military problems. In war, with Vietnam as the most glaring example, the U.S. effort has often fallen short owing to inefficient and bureaucratic command structures. In peace, America has not been able to match strategic plans to actual capabilities because of bureaucratic problems in planning. The Pentagon suffers from an abundance of interservice competition over resources and from a lack of cooperation toward a united and feasible military strategy.

This indictment of the Pentagon system led to a set of suggested reforms. The most commonly proposed remedy was to strengthen the chair of the JCS and the cross-service command structures. Where decisions had been made by consensus, the reformers proposed the chair should make them based on the advice of the service chiefs. The chair would be the principal military advisor to the president. The Joint Staff, who serve the Joint Chiefs, would be under his control and not responsible to the Joint Chiefs as a body. The chair would have a deputy chair to help run the business of the Joint Chiefs. In addition, retirement incentives should be restructured to induce longer terms of service. Interservice educational opportunities, such as the National Defense University and cross-service tours of duty, should be increased.

In general, Jones and his allies sought organizational solutions to long-term problems in military planning and action. The debate they initiated led to congressional hearings and legislative efforts to implement the bulk of their proposals, as we shall see in Chapter 7. This aspect of reform attracted much attention in part because it avoids thorny budgetary problems. Although Weinberger and the administration resisted reorganization, they preferred it to more sweeping legislation.

The push for military reform began to reach a wider audience in 1981 with the publication of James Fallows's *National Defense*, which became a bestseller.[14] Less concerned with particular ends of U.S. foreign policy, Fallows concentrates on showing how the military establishment has lost touch with common sense, if not reality. As a

---

Times Magazine, 7 November 1982; Edward C. Meyer, "The JCS," *Armed Forces Journal* 119 (April 1982): 82–90. For a longer analysis that embodies most of the concerns and proposals of Jones and Meyer see Archie D. Barrett, *Reappraising Defense Organization* (Washington, D.C., 1983), an analysis based on the DOD Defense Organization Study of 1977–80.

14. Fallows, *National Defense*.

result he advocates neither arms reductions nor rearmament, simply restructuring. Although he devotes a chapter to nuclear strategy, Fallows devotes most of his attention to conventional weapons, drawing upon the military reformers' arguments about strategy, weaponry, and management. He adopts the neutral political position that war is terrible, but we might have to fight, so we had better have the right stuff. He chides both the left and the right for simplistic arguments that cloud the debate on national security priorities.[15] Fallows captures the middle ground by arguing that we are endangering our citizens and our soldiers by building and managing the wrong kind of military. It is this pragmatic and centrist approach that was adopted by the majority of military reformers, especially those in Congress. Fallows's book was followed eventually by a host of others, which appeared with increasing frequency as the issue of reform gained popularity, especially by 1984 and 1985. Written by liberals and conservatives, these volumes tended to bemoan various aspects of the "military incompetence" of the "straw giant."[16]

The first member of Congress to take up the call for military reform was Senator Gary Hart. During the changing of the guard from Carter to Reagan, Gary Hart attempted to add military reform to the agenda of national politics. In the 23 January 1981 issue of the *Wall Street Journal*, Hart had a piece entitled "The Case for Military Reform."[17] In it, he forecast that Reagan's mandate for vastly enlarged military budgets would not necessarily translate into greater security. Hart allied himself with the growing debate he said was initiated by academics, officers, and a few politicians. This military reform movement, he said, "means three basic changes: It means spending more, selectively, for defense. It means allocating funds to

---

15. Ibid., pp. 176–84.
16. Richard N. Luttwak, *The Pentagon and the Art of War* (New York, 1985); Richard A. Gabriel, *Military Incompetence* (New York, 1985); Arthur T. Hadley, *The Straw Giant—Triumph and Failure* (New York, 1984); Richard Stubbing, *The Defense Game* (New York, 1986); Richard Halloran, *To Arm a Nation* (New York, 1986); and James Coates and Michael Killian, *Heavy Losses* (New York, 1985).
17. Hart's involvement with military reform did not begin with the op-ed pieces of the early 1980s, but the fundamental division is between what he wrote before and after the elections of 1980. Hart's entry into defense reform came in the late 1970s, when he urged recently retired Senator Robert Taft, Jr., R-Ohio, to reissue his 1976 White Paper on Defense. The actual author of the study, a Taft staffer named William S. Lind, joined Hart's office after Taft's retirement from the Senate. See Robert Taft, Jr., *White Paper on Defense*, 1978 ed. (Washington, D.C., 15 May 1978). The 1978 edition of the *White Paper* argues for the adoption of a maritime defense strategy centered on a massive shipbuilding program.

innovative weapons and programs. It means addressing a number of non-budgetary problems that, although not related to defense spending, relate directly to winning or losing wars. This includes reexamination of basic defense doctrine and concepts."[18]

Hart criticizes military practices without discussing the political purposes they serve. His assumption is that strategy (whom to fight and when to fight them) is independent of military reform (how to fight them). This very narrow conception of strategy allowed reformers such as Hart to dodge the sticky issues of the ends of U.S. foreign policy. He advocates neither cuts nor a spending binge. He consciously ignores the focus of the ensuing conflict between the peace movement and the Reagan regime: nuclear weapons and strategy. Instead, he focuses on the need for innovation in conventional strategy and weaponry, through the adoption of manuever warfare, simple and effective weapons, and a return to military leadership and education rather than bureaucratic standard operating procedures. Furthermore, Hart specifically seeks to bolster those conventional forces, the marines and mobile army divisions, that are most useful for Third World intervention—a position one would not expect of someone with Hart's credentials on foreign policy.[19] (His first role on the national political stage was, after all, as national director for George McGovern's 1972 presidential campaign, a central theme of which was ending the war in Vietnam. With his election to the Senate in 1974 as a member of the Watergate generation of Democrats, Hart quickly established his liberal credentials. He sat on the Armed Services committee and focused most of his attention there. His actions and votes consistently earned him 80 percent or higher liberal ratings on economic, social, and foreign policy issues from the *National Journal*.) Yet Hart argues that military reform "offers a new basis for something we lost in Vietnam—a genuine national consensus on defense. There is nothing ideological about the issue. It is a task in which liberals and conservatives can join."[20]

Indeed, as the clash between the peace movement and the Reagan

---

18. Gary Hart, "The Case for Military Reform," *Wall Street Journal*, 23 January 1981, p. 20.
19. For more on all these ideas see Gary Hart and William S. Lind, *America Can Win* (Bethesda, Md., 1986), in which Hart (then preparing for the 1988 campaign) and Lind state directly that they will not talk about nuclear weapons and strategy because nuclear weapons cannot be used to defend national interests, nor will they discuss geopolitical priorities.
20. Gary Hart, "The Case for Military Reform."

The Origins of Military Reform

administration engulfed Congress, more and more liberals and conservatives began to see the merit in Hart's position. While one cannot prove what influenced Hart's thinking, the evolution of his views from the late 1970s to the early 1980s suggests that the political climate pushed him in the direction of ideas that avoided the nuclear debate and general critiques of foreign policy, that carved out the uncontested terrain of conventional defense, and that chastised American fixation on high-tech weaponry.[21]

The Pentagon insiders, the authors such as Fallows, and politician Hart all found endemic flaws in the organization, leadership, strategy, and weaponry of the U.S. military establishment. Their ideas and critiques would nourish the reform movement. But intellectual ammunition alone could not mobilize a movement and make military reform a national issue. They needed a little help.

The Political Mobilization of Military Reform

Congress and the Military Reform Caucus

In 1981, Gary Hart, along with Representative G. William Whitehurst, a Republican from Virginia, founded the Congressional Military Reform Caucus (MRC) and enlisted a diverse and bipartisan membership of about fifty representatives and senators over the course of 1981 and 1982. Hart and Whitehurst found common ground on the more structural and conceptual issues of strategy and its relationship to force structure, which had been at the core of the intellectual origins of military reform. Under their leadership, the caucus focused on such matters, especially the implications of manuever doctrine for conventional warfare and weapons. During this period, the caucus served primarily as an educational forum, and its meetings were more oriented toward discussion than action.[22]

As Hart predicted, military reform attracted both liberals and conservatives. The caucus was not simply a magnet for critics and foes of the Pentagon; many prominent traditional foreign policy liberals

---

21. For more on the changes in Hart's thinking see Daniel Wirls, "Defense as Domestic Politics: National Security Policy and Political Power in the 1980s" (Ph.D. diss., Cornell University, 1988), pp. 141–47.
22. Denny Smith said that discussions "got to be a little ethereal"; Jim Courter thought "one of the flaws . . . was that it [the caucus] was only involved in discussion"; both quoted in David C. Morrison, "Caucusing for Reform," *National Journal*, 28 June 1986, p. 1597.

did not join. Although the House membership included some staunch liberals, most of the vocal advocates of the freeze were not members. Among the missing senators were Kennedy, Cranston, and Hatfield. Instead, the MRC featured some of the rising stars among the "new" Democrats, including Albert Gore, George Mitchell, and David Pryor, in addition to Hart. Some of the caucus's most visible and active members were conservatives like Representative Denny Smith, a Republican and former fighter pilot from Oregon, who led the fight against the Sergeant York (or DIVAD) gun, and Republican Jim Courter of New Jersey. Republican senatorial members also came in the "new" variety, including Charles Grassley, Nancy Kassebaum, and Mark Andrews, three active members from midwestern farming states. The MRC drew many members caught between forces in favor of a military buildup and forces that did not want to give the Pentagon whatever it wanted while ignoring the deficit and domestic programs. As Denny Smith put it, "I don't want them hanging $600 toilet seats around my party's neck."[23] Or as conservative Alabama Democrat Bill Nichols said, "Every time I go home, people grab me by the shoulder and say, 'Hey, I'm for defense, but . . . .'"[24] Because reformers came in different ideological stripes, motivations ranged from a conservative wish to rebuild the Pentagon's image and maintain the buildup, to a moderate inclination to balance competing political pressures, to a liberal intention to cut the buildup from behind a mask of reformist credibility. Overall, the caucus would come to mirror the ideological diversity of Congress itself (Table 7). Reformers exhibited a considerable degree of convergence in viewpoint on general principles. Among the areas of agreement were emphases on conventional over nuclear weapons, maneuver over attrition doctrines, simpler and more numerous weapons, and increased resources for nonprocurement categories. The mixture of incompatible ideologies and motivations, however, prevented the MRC as a whole from uniting behind many particular issues or applications of reform ideas. Nevertheless, individuals and small groups within the membership sponsored most of the reform legislation discussed below and in Chapter 7.

Because the material reality of the Reagan military budgets required more than discussion, some members of the caucus began to

---

23. Quoted in *Business Week*, "An Unlikely Alliance Takes on the Pentagon," 5 August 1985, p. 54.
24. Quoted in *Congressional Quarterly Weekly Report*, 11 May 1985, p. 887.

The Origins of Military Reform

Table 7. Ideological diversity of the Military Reform Caucus compared to the Ninety-ninth Congress (in percentages)

ADA ratings	MRC	Full chamber	Chamber excluding MRC
	House MRC (50 Democrats and 48 Republicans)		
0–25	40 (39)	40	40
26–50	10 (10)	14	15
51–75	24 (24)	22	22
76–100	26 (25)	23	23
Average	46	45	45
	Senate MRC (16 Democrats and 11 Republicans)		
0–25	26 (7)	46	53
26–50	22 (6)	15	15
51–75	26 (7)	18	15
76–100	26 (7)	21	19
Average	50	40	37

Sources: Military Reform Caucus membership list for February 1986; Americans for Democratic Action (ADA) voting scores for 1985, reprinted in *Congressional Quarterly Weekly Report*, 22 November 1986, pp. 2959–66.

Note: Numbers in parentheses are the actual numbers of MRC members in each quartile. The same distribution obtains when the ratings on arms control and on nuclear weapons by the Council on a Livable World are treated accordingly.

take the principles of reform and apply them legislatively. The first amendments to come from the reformers were on conventional tactics, training, and leadership rather than waste, fraud, and abuse in contracting, which would later become the focus of reform. For example, in May 1982 several amendments were appended to the FY 1983 defense authorization bill. Gary Hart's amendment required the secretaries of the army and navy to report on the extent to which manuever doctrine had been incorporated into policy and training. Slade Gorton's required all the services to report on the extent of military history education in the military schools. Sam Nunn's had the secretary of defense report on the National Guard's Vista 1999 study, which endorsed maneuver tactics. Finally, John Warner's amendment asked the defense secretary to report on initiatives to increase unit cohesion. These were the issues that were the heart of the MRC's agenda under Hart and Whitehurst. The problem was that Congress could not legislate tactics or good military thinking. Most of the legislative efforts were educational or exhortatory. And, more important, they did not attract many adherents to the cause of military reform. Moreover, several caucus-sponsored initiatives that dealt with specific weapons were defeated in 1981 and 1982. For example, amendments to get diesel submarines for the navy and motorcycles for the army failed, as did attempts to cut appropriations for DIVAD and the M-1 tank. Some of these failures were reflections of divisions within the caucus. While Hart favored scrapping plans to build more large nuclear aircraft carriers and, instead, building smaller carriers, William Whitehurst, the caucus cochair, opposed this. Virginia, Whitehurst's state, was where the large carriers were being built.[25] These specific initiatives pertaining to force structure and weapons were often internally divisive, relatively unpopular, and initially unsuccessful. However persuasive the reform arguments were, they were not at first politically powerful.

Exposing Waste, Fraud, and Abuse in the Pentagon: The Reform Triangle

The political power of military reform grew and spread only with a change in focus from the more intellectual issues of tactics, leadership, and force structure to issues of the waste, fraud, and abuse in the military procurement process. This transformation sprung from an alliance among public interest groups, and the media, and members of Congress—plus some scandalous information about the way the Pentagon was spending tax dollars. In political analysis the term *iron triangle* usually refers to a nexus of interest and cooperation among a bureaucratic agency, a congressional committee, and some private interest, usually in industry or business. These triangles are characterized as "sub-governments" because they form agenda-setting and decision-making units that exclude broader participation. The Pentagon and the relationships of its subdivisions with contractors and defense committees have been the frequent objects of this kind of analysis.[26] Partly in response to these iron triangles, outsiders have mobilized reform triangles.[27] Unlike iron triangles, which thrive on secrecy and limited participation, reform triangles seek to

---

25. "Congress' Military Reform Caucus: Slow to Start on New Defense Initiatives," *Armed Forces Journal International* (April 1982): 17.

26. Gordon Adams, *The Politics of Defense Contracting* (New Brunswick, N.J., 1982).

27. This type of triangle was critical to the success of the environmental and consumer movements of the 1960s and 1970s. Paul H. Weaver describes similar relationships of regulatory agencies, public interest groups, and allies in Congress and the courts as the "new iron triangles"; see his "Regulation, Social Policy, and Class Conflict," *Public Interest* 50 (Winter 1978): 52.

The Origins of Military Reform

advance their causes by expanding the scope of conflict over policy. The mechanism depends on mutual interests among public interest groups, the media, and Congress. In the case of military reform, information about problematic Pentagon programs, often leaked by disillusioned or disgruntled employees of the bureaucracy or the private sector (called "whistle-blowers"), is publicized by the media, sometimes with the help of a public interest group. The resulting publicity prompts congressional hearings and investigations, initiated by members of Congress either out of sincere concern or a desire to use the cause to advance their own political interests. The process often culminates in legislative remedies.

Just as authors of different political stripes took up the cause of defense reform, so did ideologically diverse public interest groups. Several of the institutions that devoted resources to promulgating military reform preceded the 1980s movement itself.[28] One organization in particular, however, had its roots in the 1980s and exemplified the pragmatic and putatively nonpartisan approach of James Fallows and Gary Hart. The Project on Military Procurement was founded in 1981 as a project of the National Taxpayers Legal Fund (NTLF). The founder, Dina Rasor, had been a researcher at the National Taxpayers Union, a group with conservative-libertarian philosophy and leadership. When the union would not sponsor Rasor's project, A. Ernest Fitzgerald, Rasor's mentor and one of the most famous Pentagon whistle-blowers, helped get the NTLF to adopt her project. Later, when libertarian and antimilitary liberals on the board of NTLF objected to the project's apolitical stance on foreign policy, the project was dropped by that group and picked up by the Fund for Constitutional Government, a public interest group that seeks to expose fraud and waste in government.[29] According to its own literature the goals of the Project on Military Procurement are: "to reform

---

28. On the left, the Council on Economic Priorities has for years produced critiques of the defense establishment based on quantitative testing and economic data and consistent with the arguments of the movement. As its name implies, the council has well-defined priorities: it would like to see money taken from the military and given to domestic programs. Other organizations, such as the conservative Heritage Foundation, took up the cause only as it became popular. The foundation produced studies that endorse many of the central criticisms and proposals of reformers.

29. Dina Rasor, *The Pentagon Underground* (New York, 1985), chaps. 2–4. The Fund for Constitutional Government, which is a liberal group founded and funded by Stewart Mott, provided about one-quarter of the project's funding. The rest came from foundations, including the Rockefeller Family Fund and the Reuben Fund, and other private donations.

the Pentagon procurement system by educating the public and the Congress of the ongoing waste, fraud and abuse through the press; to provide an effective and reliable defense while saving the American taxpayer as much money as possible; and to assist in any way, whistleblowers in the military establishment who wish to expose those abuses."[30] The project maintained a studied silence on foreign and defense policy because it did not want to be ignored or attacked on the basis of political affiliations or biases.[31] It avoided nuclear weapons, ostensibly because testing data are either hard to obtain or unavailable. It was strictly nonpartisan. Like many entities in the reform movement, the project only wanted to talk about the means of national security, not the ends.

Methodologically, the project concentrated on "closet patriots" in the form of defense establishment insiders, either in government or industry, who want to leak information on DOD projects.[32] It served primarily as a conduit between these whistle-blowers and the press, providing the media with documented inside information and protecting the insider while giving the media a more "legitimate" source to cite and blame, should trouble arise. This allowed the press to report on the military's problems without appearing to be biased. The project also arranged for congressional briefings on their research. Thus the project and the press together provided Congress with information and publicity crucial to its effort to establish the middle ground between the peace movement and the administration.

The media flocked to the kind of stories and controversies leaked by insiders and publicized by the Project on Military Procurement and Congress. The reasons for this are several. First, the split between the cold war coalition and the peace movement put the press in an awkward position. Reporting on either or both, they ran the constant danger of the appearance of bias. The debate was very ideological, and the media do not handle ideological issues adroitly or comfortably. They are more comfortable with apparently objective information and arguments rooted in the logic of money and common sense.[33] One only need contrast the emotional uproar over "The Day After" with the sober indignation of a "60 Minutes" or "20/20" exposé of a Pentagon boondoggle to get an appreciation of the dif-

---

30. Project on Military Procurement, "Who We Are, What We Do, Why We Do It" (n.d.), p. 1.
31. Rasor, *Pentagon Underground*, pp. 92, 101, 105–6.
32. Project on Military Procurement, "Who We Are," p. 1.
33. Rasor, *Pentagon Underground*, p. 106, also makes this point.

ference. When reformers began to provide the media with stories of weapons that did not work and seven hundred–dollar toilet seats, the media were on solid ground. Who could be for waste, fraud, and mismanagement? They had an unassailable position from which to attack the Pentagon. The major Pentagon correspondents produced a steady and increasing stream of stories for both TV and print journalism. Representatives and senators find national press coverage from respectable sources, like the *Wall Street Journal* and the *New York Times*, hard to ignore and easy to exploit.

The first task the Project on Military Procurement undertook, in 1981, was the attack on the M-1 Abrams battle tank, built at the time by Chrysler. Dina Rasor fed the media inside information on the tank which detailed rapidly rising costs and various defects such as fragile steel treads and finicky turbine engines. A prominent series of articles and reports resulted, including coverage by the *Wall Street Journal*. Although a few members of Congress showed some interest in pursuing the tank's problems, the political importance of military reform had not yet struck the legislature. The army waged an extensive public relations campaign to salvage the tank's reputation, and the program easily survived the assault.[34] Soon, however, another scandal developed that Pentagon public relations could not manage, a scandal that brought military reform to national attention and revealed how politically powerful military reform could be.

The Spare Parts Scandal: Bringing Military Reform to the Public

The growing military reform coalition discovered that the smallest details of procurement—the costs of individual weapons components and spare parts—carried the potential for powerful publicity. As several leaders in the movement stated, most citizens are not sure what a tank or bomber should cost, but they do know that a toilet seat should not cost $700 dollars nor a claw hammer $435. The exposure of parts-pricing scandals, perhaps more than any other issue, brought military reform to the attention of the American public.[35]

The reform triangle revealed and publicized the spare parts scandal, which received its impetus in the summer of 1982 from a leaked

---

34. On the M-1 campaign see Rasor, ibid.; and Goodwin, *Brotherhood of Arms*, p. 251.
35. Rasor, *Pentagon Underground*, pp. 148 and 288, credits her underground sources for the idea of exposing parts scandals as a way to reach the public.

memo authored by an air force official, criticizing Pratt and Whitney for exorbitant price increases.[36] When the Project on Military Procurement released the memo to the press, a flood of stories followed along with numerous requests to the project for more spare parts stories. The Pentagon only succeeded in contributing to the scandal by explaining that the increases were "justifiable" under its pricing procedures. In March 1983, ABC's "20/20" aired a report on parts overpricing, which provoked rebuttals from the Pentagon and White House and inquiries and outrage from Congress.

The program led to other revelations, including the $435 claw hammer the navy purchased from the Gould Corporation and the $1,000-plus stool cap bought for the AWACS aircraft. Congressional investigations soon followed. Bill Nichols's House Armed Services Investigations subcommittee held hearings over several months in 1983 which focused on the Pratt and Whitney problems. The Senate Armed Services committee also held hearings that fall. Later in the fall, the House Committee on Government Operations released a report charging that the Pentagon's recent attempts to reform spare parts procurement had failed. The revelations culminated in legislation being introduced during the 1984 congressional session. Virtually identical bills in both chambers sought to compel the Pentagon to use more competition, sign multiyear contracts, buy commercial parts, and get technical data for spare parts. The bills opened with the invocation "Congress finds that recent disclosures of excessive payments by the Department of Defense for spare parts have undermined confidence by the public and Congress in the defense procurement system."[37]

---

36. Ibid., pp. 160–62; A. Ernest Fitzgerald, *The Pentagonists* (Boston, 1989), pp. 150–55. This was the so-called Hancock memo. Robert Hancock, an official at Tinker Air Force Base in Oklahoma, wrote the memo accusing the major jet engine contractor of thirty-four spare parts price increases of 300 percent in one year.

37. For the Nichols hearings see U.S. House, Committee on Armed Services, Subcommittee on Investigations, *Examination of Armed Services Policies and Procedures in the Procurement of Spare and Repair Parts, and the Pricing Thereof of These Items*, 98th Cong., 2d sess., 19 and 20 April, 25 May, 9 June, 10 October 1983. For the Senate hearings see U.S. Senate, Committee on Armed Services, *Spare Parts Procurement for the Department of Defense*, 98th Cong., 1st sess., 26–27 October 1983. Chairman John Tower's and Ranking Minority Member Sam Nunn's opening remarks both cited the recent Pentagon spare parts "horror stories," including the $435 claw hammer and the nearly $1,000 stool cap. Tower's first point was that the examples had been revealed by Pentagon personnel working within the system; his second point was that the administration was not to blame for problems that had been going on for decades. The House report is *Failure to Implement Effectively the Defense Department's High Dollar*

The Origins of Military Reform 99

Continual, if somewhat exaggerated, claims by Weinberger and other Pentagon officials that spare parts revelations were the work of a vigilant Defense Department fell largely on deaf ears. Other officials tried to squelch the scandal by arguing that the whole affair was due to a misunderstanding of Pentagon accounting procedures. They argued that a policy called "equal allocation of overhead" distributed an equal percentage of overhead, or indirect costs of production, across all components of production. Thus a $10 hammer might have $425 in overhead added to its price for accounting purposes. It turned out, however, that the Pentagon and its contractors did not use the equal allocation procedure, and accounting did not explain the prices.[38] It mattered little who discovered the scandals or what the exact cause was. With the budget crunch, the inherent irrationality of the pricing scams was intensified. Weinberger had to act. First, in August 1983 he issued a ten-point directive for the procurement of spare parts. Whatever effect the directive had on prices, it did not quell the uproar. In December 1984, he had to create a deputy assistant secretary of defense for spare program management, the first of several such appointments he was to make in response to congressional criticism.[39]

Weinberger's remonstrations notwithstanding, whistle-blowing efforts aided by the Project on Military Procurement and the media continued to keep the spare parts scandal growing as the 1984 election season approached. During that summer, Rasor gathered data from two air force employees on parts prices in the C-5, including the $7,600 coffee pot and the $670 arm rest. The air force had ignored the employees' suggestions about the arm rest and then stonewalled once the prices became public. One avenue for the publicity was the hearings held in September 1984 by MRC member

---

Spare Parts Breakout Program Is Costly, Fifteenth Report by the Committee on Government Operations, 11 November 1983. This committee, chaired by Jack Brooks (D-Tex.), played a major role in several reform investigations. The spare parts bills were H.R. 5064 and S. 2572. For the House version see U.S. House, Committee on Armed Services, Subcommittee on Investigations, *Hearings on HR 5064, Defense Spare Parts Procurement Reform Act and HR 4842*, 98th Cong., 2d sess., 13 and 21 March 1984.

38. For a thorough analysis of the spare parts scandal and the chimerical "equal allocation" procedure see Fitzgerald, *Pentagonists*, chaps. 9, 10, and 12.

39. For Weinberger's own account of his spare parts reform initiatives and their impact see his testimony and submission of the Executive Summary of the FY 1986 Annual Report in U.S. House, Committee on Appropriations, Subcommittee on Defense, *Department of Defense Appropriations for 1986*, 99th Cong., 1st sess., 1:48–49, 150–52, 384–86.

Charles Grassley through his Judiciary Subcommittee on Administrative Practices and Procedures. Another was an NBC "Today" show that featured the two employees. By this time, the list of spare parts with astronomical price tags had become quite lengthy, and more examples were not hard to come by.[40]

Although the presidential campaign diverted attention from such stories, by fall 1984 many people would have concurred with Democratic representative Charles Schumer's assessment: "What welfare mothers did for social spending in the 1970s, $6,000 coffee pots are doing for defense spending in the 1980s."[41] The spare parts scandals provided political ammunition for committed reformers and recent converts. It made the issue of reform readily accessible and the need for reform quite obvious. More important, it pointed the way toward politically effective and popular reform. The inexcusable and attention-grabbing waste, fraud, and abuse, although hardly new to the Pentagon, provided just the additional political leverage the military reform movement had heretofore lacked to promote its program of weapons simplification, procurement reform, and economic efficiency in defense expenditures. Moreover, as the Reagan administration pressed for record military budgets, the peace movement kept public opinion and many politicians wary of unrestrained militarism. Budget deficits became one of the leading issues on the political agenda at the same time that defense budgets produced the enormous cost overruns and contract abuses. This mixture of circumstances could only contribute to pressure for congressional support of military reform. But congressional interest in reform did not blossom overnight. Although the membership and influence of the MRC increased each year, the caucus only grew from about fifty to eighty members during 1983 and 1984. Despite the auspicious publicity generated by pricing scandals, the caucus and the movement as a whole were overshadowed through 1983 by the fight over nuclear weapons and arms control; they did not capture and hold center stage until the peace movement had crested and the 1984 elections were history, when both Democrats and Republicans were searching for new ideas. From then until the 1986 elections, military

---

40. Rasor, *Pentagon Underground*, pp. 148–60. Other examples included a modified $7/64$-inch Allen wrench the air force bought for $9,609 a piece, which was worth less than a dollar commercially; a 4-cent diode purchased for $110; and standard machine screws for $37 each.

41. Quoted in *Congressional Quarterly Weekly Report*, 14 September 1985, p. 1799.

reform and members of the caucus figured prominently in the defense debate. With the waning of the peace movement and the issues of nuclear war and arms control, military reform would become the basis for the next wave of conflict between the administration and members of Congress over the military budget and programs.

5

The Nuclear Freeze Campaign and the Democratic Party

The political counterattack to the Reagan revolution began when the freeze campaign reached national prominence. Although Washington was the ultimate target of its war for public opinion, the peace movement leadership wanted to continue building public support before taking their battle to to the capital. But politicians in Washington needed the freeze; for no other issue seemed endowed with so much potential for blunting, if not reversing, the Reagan agenda. Local pressure and national polls were rapidly converting members of both houses of Congress to the cause. The White House grew anxious, while Democratic leaders, especially in Congress, grew hopeful with anticipation. Some Democrats were ready to lead the charge on Capitol Hill whether the freeze movement had its legions in battle formation or not.

Legislating Arms Control Policy

The Freeze in Congress

The first politicians to hear and heed a call for leadership were Senator Ted Kennedy and Democratic representative Ed Markey of Massachusetts. Kennedy's presidential ambitions had him on the prowl for issues. Out on the hustings in late 1981 and early 1982, Kennedy sensed the importance and the potential of the peace movement.¹ Not only did the issue of arms control and the freeze seem like a winner; it also would give him, as a presidential candidate, credibility on foreign policy. Kennedy linked up with dovish Republican senator Mark Hatfield of Oregon and began to draft a freeze resolution for congressional action. The Kennedy-Hatfield partnership was a fortuitous alliance for Representative Markey, who had been designing a resolution and seeking support but needed an ally with national recognition.² Joint resolutions in favor of a freeze were introduced to both houses in March 1982. Many within the leadership of the freeze campaign considered this legislative initiative premature; it was well in advance of the timetable they had envisioned. They wanted a bipartisan block of at least 150 cosponsors, as opposed to Markey's mostly Democratic 42, and believed some of the resolution's wording was inappropriate.³ The Senate version (S.J. Res. 163) had only 18 sponsors. With a Republican majority in the Senate and a solid conservative majority in the House, both resolutions faced uphill battles. Nevertheless, the House version (H.J. Res. 521) quickly gathered 176 cosponsors, and on 23 June emerged relatively unscathed from the Foreign Affairs committee on a solid 26–11 vote, with 7 GOP committee members in favor.

Despite the size and omnipresence of the movement, and the strength of the committee vote, the freeze fell just short on the house floor. The freeze resolution, known as the Zablocki resolution for its sponsor, Clement Zablocki, finally came to a vote in the House in early August after a lengthy and often confused floor debate,⁴ which often devoted more energy to the wisdom or folly of Reagan defense and arms control policies in general than to the merits of the freeze in particular. In fact, a negotiated freeze was only one of the arms control objectives urged in the Zablocki resolution. Floor amendments further diluted any pointed reference to an immediate freeze as "the overriding objective" with clauses and subsections on verification, essential equivalence, reductions, and de-

---

1. For Kennedy's description of how the grassroots movement gave impetus to his involvement see Edward M. Kennedy and Mark O. Hatfield, *Freeze! How You Can Help Prevent Nuclear War* (New York, 1982), pp. 123–24.
2. Douglas C. Waller, *Congress and the Nuclear Freeze* (Amherst, Mass., 1987), chap. 3. Waller was Markey's aide in charge of the freeze resolution, and the book provides the most detailed history of the freeze resolution in Congress.
3. Markey's resolution was H.J. Res. 433. Waller, *Congress and the Nuclear Freeze*, pp. 58–59.
4. *Congressional Record*, 5 August 1982.

stabilizing weapons. Nevertheless, the White House took the debate and vote quite seriously and mounted a substantial lobbying effort, including briefings by Weinberger and Secretary of State George Shultz and phone calls to wavering representatives from Reagan's arms negotiators in Geneva, Edward Rowny and Paul Nitze. The modified resolution endorsed by the movement lost on a 204–202 vote to a Reagan-backed substitute that called for negotiated reductions to be followed by a freeze.[5] The substitute was essentially an open-ended endorsement of Reagan's Strategic Arms Reduction Talks (START) proposal. Both sides claimed victory.

The barely victorious substitute resolution had originated in March when the administration and its allies began their counterattack against the growing influence of the peace movement. Just two weeks after the Kennedy and Markey resolutions were introduced, senators Scoop Jackson and John Warner presented S.J. Res. 177 with 56 cosponsors, including the Democratic minority leader. A parallel resolution was submitted to the House. Both substituted the simple but crucial phrasing: "long-term, mutual and verifiable nuclear forces freeze at equal and sharply reduced levels of forces."[6] Here the word "forces" negated the freeze's impact on testing and production. "Equal and sharply reduced" meant that, before any freeze, negotiators would have to implement a major arms reduction treaty—a treaty that would do what no other has ever done, namely, reduce forces to levels both sides agreed were equal. The whole thrust of the freeze was of course to stop now and avoid negotiations over ratios and other details. Reagan immediately approved of the Jackson-Warner resolution in a press conference, during which he warned, "On balance, the Soviet Union does have a definite margin of superiority."[7] Two weeks later, in a speech in Tennessee, the president argued that a freeze "would legitimize the position of great advantage for the Soviets" and that "we must go beyond a freeze to get at real reductions of nuclear arms."[8]

Reagan turned this rhetoric into policy in early May, when he un-

---

5. The substitute language was sponsored by William Broomfield (R-Mich.). On the debate see Edward Feighan, "The Freeze in Congress," in *The Nuclear Freeze Debate*, ed. Paul M. Cole and William J. Taylor, Jr. (Boulder, Colo., 1983), pp. 34–39.

6. The Jackson-Warner resolution is reprinted in U.S. Senate, Committee on Foreign Relations, *Nuclear Arms Reduction Proposals*, 97th Cong., 1st sess., 14 April and 18 May 1983, pp. 192–93.

7. Press conference text reprinted in *Congressional Quarterly Weekly Report*, 3 April 1982, p. 767.

8. Ibid., 20 April 1982, p. 640.

veiled his START proposal just as the arms control and freeze debate began in Congress. The proposal ignored any notion of a freeze and focused on significant reductions in strategic missiles and missile warheads. The administration made repeated suggestions that congressional approval of the freeze or any reductions in the strategic buildup would jeopardize the American position in START talks in Geneva. Reflecting the irony of the administration's sudden fixation on disarmament, freeze advocates were forced to criticize Reagan's reduction proposals and defend the primacy of a moratorium. The often-made argument that support for a freeze would undermine the president's position in Geneva received much of the credit for the close vote in August. The freeze never made it through committee in the Republican Senate.

The Antifreeze Campaign

Neither the defeat of the freeze in Congress nor the START initiative quieted attacks by New Right organizations and the administration against the still-growing peace movement and the freeze. The New Right coalition of organizations, which had so successfully raised the specter of military inferiority in 1980, reacted to the power of the freeze with an "antifreeze" campaign. Member organizations of the Peace through Strength Coalition, especially the American Conservative Union and the National Forum, initiated a mass media program to revive the peace through strength theme and associate the peace movement with Soviet foreign policy goals.[9] The October 1982 issue of *Reader's Digest*, the most widely read magazine in America, featured senior editor John Barron's "The KGB's Magical War for Peace." Barron argued that the KGB and its communist allies in America greatly influenced, if not controlled, the freeze movement. Soon after the article's publication, President Reagan remarked to an audience in Columbus, Ohio, that the freeze campaign was "inspired not by the sincere, honest people who want peace, but by some who want the weakening of America."[10] Just before Reagan's Columbus speech, Senator Jeremiah Denton had the *Digest*

---

9. David Corn, "Antifreeze Campaign Heats Up," *Nuclear Times*, February 1983, pp. 6–7.
10. Quoted in George Lardner, Jr., "Soviet Role in Nuclear Freeze Limited, FBI Says," *Washington Post*, 26 March 1983, p. 7. See Robert Freidman, "Rees to Barron to Reagan," *Nuclear Times*, November/December 1982, p. 25, for the origins of Barron's arguments.

article read into the *Congressional Record* to substantiate his charge that Peace Links was manipulated by communists. Peace Links, headed by Betty Bumpers, wife of Senator Dale Bumpers, was an organization of mostly white, middle-class women predominantly from the South and West. Barron does not refer to the group in his article. Outrage from Bumpers and others compelled Denton to reconsider his accusations. Charges of communist leadership and manipulation and countercharges of red-baiting led to the public release in December 1982 of the details of a House Intelligence committee secret hearing into the matter. The hearing, at least as reported in the press, minimized the involvement and impact of Soviet "active measures" and American communists in the movement.[11] Although the freeze withstood the red-baiting more successfully than had its predecessor peace movement in the late 1950s, the administration's "active measures" against the freeze drew attention away from the substantive debate.

The attacks by the administration and its allies played consciously upon the ambivalence and polarities in public opinion. While public opinion heavily supported a freeze and arms control, Reagan was elected partly on the basis of his proposed defense policies. Public opinion was anything but univocal, unambiguous, or, above all else, stable. Support for the freeze, specifically, and for the other goals of the peace movement proved significantly weaker than some polls implied. It all depended on the exact questions asked. Analysis of a Gallup-run poll from late October and early November 1982 demonstrates this point.[12] Of the 79 percent of those polled who were in favor of a freeze, 21 percent thought the United States should freeze only after completing the rearmament process, a position supported by Reagan and his Republican allies (Table 8). Support for an immediate freeze dropped to 58 percent of all voters and to 60 percent of those respondents (who identified themselves as Democrats or Inde-

---

11. The most influential information came from the FBI. See Lardner, "Soviet Role"; and Leslie Maitland, "FBI Rules Out Russian Control of Freeze Drive," *New York Times*, 26 March 1983, p. A1, 7.

12. Unless otherwise noted, the data in this section comes from analysis of the Chicago Council on Foreign Relations data set, *American Public Opinion and U.S. Foreign Policy, 1982* (Ann Arbor, Mich., 1983). October and November 1982, when the poll was taken, was perhaps the low tide of the first Reagan administration. The recession and the peace movement had combined to drop the popularity of both the president and his party, probably causing this poll to measure more support for the freeze and Democratic sentiments than it might have at many other periods during these years.

Table 8. Qualified support for a freeze (in percentages)

	In favor of a freeze	Now	After buildup
All respondents	79	58	21
Democrats and those leaning Democratic	80 (44)	60 (33)	20 (11)

Source: Chicago Council on Foreign Relations, *American Public Opinion and U.S. Foreign Policy, 1982* (Ann Arbor, Mich.: Inter-University Consortium for Political and Social Research, 1983).
Note: Numbers in parentheses are percentage of all respondents.

pendents leaning toward the Democrats) likely to support Democratic candidates.[13]

A scale of support for the peace movement's goals, which included an immediate bilateral freeze, cuts in the defense budget, and a willingness to halt unilaterally particular strategic programs, displays the ambivalence in public opinion. The first column in Table 9 shows the percentages favoring an immediate freeze; the second column, the percentages for cutting the defense budget; the third column, those in favor of unilateral arms control. The fourth column isolates those who want both an immediate freeze and a cut in defense. The last column takes the combined profreeze and pro–

Table 9. Scale of Support for Freeze Movement Objectives (in percentages)

	Individual objectives			Combined objectives	
	Freeze now	Cut defense	Unilateral halt	Freeze and cut	Freeze, cut and unilateral halt
All respondents	58	34	19	25	10
Democrats and leaning Democratic	60 (33)	41 (23)	21 (12)	31 (17)	12 (7)

Source: Chicago Council on Foreign Relations, *American Public Opinion and U.S. Foreign Policy, 1982* (Ann Arbor, 1983).
Note: Numbers in parentheses are percentage of all respondents.

13. However much people spoke of bipartisanship on this issue, the vast majority of freeze supporters in Congress were Democrats. In the 1982 elections, the peace movement's future depended on its ability to elect profreeze Democrats. In this poll, partisan identification with the Democrats is unusually high: 43 percent identified themselves as Democrats, 23 percent as Republicans, and 31 as Independents. Again, the support for Democrats and the freeze is, if anything, exaggerated in this poll.

Table 10. Support for matching Soviet Military power (in percentages)

	Very important	Somewhat important	Not important
All respondents	50	34	12
Republicans	61	28	6
Democrats	47	35	12
Independents	45	37	15

Source: Chicago Council on Foreign Relations, *American Public Opinion and U.S. Foreign Policy, 1982* (Ann Arbor, 1983).

defense cut groups and determines how many also favor unilateral arms control by the United States. At each step, the support for the movement's goals drops significantly. Perhaps most telling is the fourth column: only 25 percent of all respondents, and 31 percent of likely Democratic voters, favor both a freeze and a cut in defense spending.

Finally, most potential voters of all persuasions believed that matching the Soviet Union's military might was a very important goal of U.S. policy (Table 10). Very few thought it unimportant. This indicates how susceptible the public was to Reagan's argument that the freeze or unilateral restraint would leave America militarily inferior to the Soviets. Other polls showed that Americans favored military superiority over the Russians and would not support a freeze if it gave the Soviets the edge, which is just what Reagan repeatedly stated it would do. For the first half of the 1980s, in fact, the majority consistently believed the Soviet Union to be superior in nuclear armaments. Finally, in some of the same polls that had 70 percent favoring a freeze, as many as one-third or more of the respondents thought that Reagan backed a freeze, while another third did not know the president's position.[14]

The public had no firm convictions, and the pollsters asked too many questions for either side to prevail decisively. This typifies the dynamics of public opinion warfare. Both sides could cite polls and interpret them to fit their needs. The peace movement relied on public opinion data to convey its power, but that same data could be used to publicize and support quite different interpretations of the general will. Reagan and his allies were fully aware of the superficial aspects of public support for the freeze and exploited them in their

---

14. *Gallup Reports* 212 (May 1983): 26–27.

efforts to undermine the movement's attempts to turn its message into public policy.

The 1982 Elections and the Freeze in 1983

Nevertheless, the results of the 1982 elections kept the initiative with the peace movement. Although few House and Senate outcomes could be directly attributed to the power of the freeze, the issue pervaded many campaigns despite the prominence of the recession. Politicians were compelled to discuss and take a stand on arms control and the strategic buildup. The elections were the culmination of the nationwide campaign that had produced endorsements of the freeze in about eight hundred town meetings, city councils, and county governments and in seventeen state legislatures. Over one hundred national organizations, including several unions and labor organizations, had signed on. On election day, in addition to numerous local votes, freeze referenda passed in eight states (including California) and the District of Columbia, losing only in Arizona. An estimated 30 percent of the U.S. electorate had the opportunity to vote on the freeze, and about 60 percent cast their ballots in the affirmative. Profreeze candidates fared well in the elections as the Democrats picked up twenty-six seats.[15]

At the February 1983 annual freeze conference in St. Louis, the theme for the year was "From Popular Mandate to Public Policy."[16] The primary goal was to translate the momentum of public pressure into congressional approval of the freeze, first in the House then in the Senate. The convention went on to endorse three new initiatives that conveyed a sense of growing confidence and power. First the delegates made the elections of 1984 a top priority, especially the election of a sympathetic president. Next the convention called for congressional suspension of "funding for the testing, production, and deployment of U.S. nuclear weapons, calling upon the Soviet Union to exercise corresponding restraint"—the so-called quick freeze. Furthermore, the delegates called for the delay of deployment of the Euro-missiles. Both measures had elements of the unilateralism the movement earlier had sought to avoid. Finally, in an

---

15. Representative Markey's office estimated that the freeze gained about twenty to thirty votes as a result of the elections; see Waller, *Congress and the Nuclear Freeze*, p. 165.
16. For coverage of the third annual convention see David Corn, "Freeze Sets Daring Strategy," *Nuclear Times*, March 1983, p. 6.

effort to expand its constituency beyond the white middle class, the convention endorsed Jobs with Peace Week (to establish links with labor) and the Martin Luther King twentieth-anniversary march on Washington. As one indication of the movement's perceived importance, all three networks had correspondents and crews in attendance, as did major papers, including the *New York Times* and *Washington Post*.

This heady and mildly radicalized atmosphere within the movement anticipated a congressional victory early in the new year. Along with fellow Massachusetts representative Silvio Conte, Markey reintroduced the freeze resolution on the first day of the Ninety-eighth Congress with 150 cosponsors. Kennedy and Hatfield did likewise in the Senate with 34 cosponsors. Zablocki reintroduced the language of the resolution passed the previous spring by the Foreign Affairs committee as the basis for committee action. After the resolution (H.J. Res. 13) sailed out of the House Foreign Affairs committee on 8 March with a 27-9 vote, the fight on the House floor began anew on 16 March, with thirteen hours of debate. Amid the inspection of every word, phrase, and nuance, freeze advocates had several times to turn away "reduce then freeze" or "freeze and/or reduce" language and fend off repeated charges of unilateralism. What had been intended as a one-day debate to culminate in a vote turned into one of the longest and most arduous debates in congressional history.[17] It was certainly the most arcane. Although by the end of the first day it was clear that the freeze had enough votes to pass, the mainly Republican opposition refused to terminate debate and vote. Instead, freeze opponents proliferated amendments in an attempt to gut the resolution and erode its support. By the conclusion of each ensuing day of debate (on 13, 20, 21, and 28 April), there were more amendments pending than when the day began. Republicans were writing amendments faster than the leadership could close debate and vote on them. The tactic amounted to what one observer described as "filibuster by amendment."[18]

The amendments ranged from serious threats to the integrity of the resolution to harmless declarations like that offered by Robert Walker (R-Pa.), which stipulated that the resolution did not "man-

---

17. For the complete floor debates see *Congressional Record*, 16 March 1983, H1201–62; 13 April, H1999–2064; 20 April, H2197–2238; 21 April, H2256–85; 28 April H2457–67; 4 May H2604–61.
18. Waller, *Congress and the Nuclear Freeze*, p. 258.

date any agreement that would jeopardize our ability to preserve freedom." An example of a serious threat was Georgia Democrat Elliot Levitas' attempt to require that bilateral reductions must follow a freeze "within a reasonable, specified period of time," with the implication that the freeze would terminate if such reductions were not obtained—what some called a freeze "sunset clause."[19] This proved to be the one significant amendment; others addressed specifications, such as whether the freeze included submarines and bombers, which were weapons platforms and not actual nuclear bombs like missile warheads. The profreeze leadership employed a largely successful "embrace or deflate" strategy whereby they accepted harmless amendments such as Walker's and amended the threatening language in serious amendments. At the close of the 28 April debate, with dozens of amendments still pending, the freeze leadership took an unprecedented step by asking the Rules committee to revise its rule for H.J. Res. 13 and set a limit on debate. The extraordinary move was agreed to in the committee and in a floor vote.[20] On 4 May, after about fifty hours of debate spread over seven weeks, the House voted 278–149 for a somewhat diluted freeze resolution.

In the end, about one hundred amendments were offered to the original resolution, and nearly thirty were accepted or passed. Although the most damaging amendments were defeated, others became part of the final language and introduced many ambiguities about the ultimate intent of the resolution. What had been about a two-page resolution, focused on the freeze, emerged from the House as seven pages, with the freeze buried amid numerous qualifications and stipulations. Finally, although a joint resolution has the force of law, the H.J. Res. 13 stated that U.S. arms control negotiations "should" (not "shall") pursue a freeze and the other objectives in the resolution.[21] There was nothing mandatory about the legislation.

With the passage of the freeze in the House, the move from popular mandate to public policy ground to a halt. The proposal, once

---

19. The Levitas amendment took up much of the last day of debate; *Congressional Record*, 4 May 1983, H2631–47; and *Congressional Quarterly Weekly Report*, 7 May 1983, pp. 868–69.

20. On the rule change and Rules chair Claude Pepper's defense of it see *Congressional Record*, 4 May 1983, H2604–5.

21. The resolution as passed by the House is reprinted in *Congressional Quarterly Almanac, 1983*, pp. 210–11.

again, went nowhere in the Republican-controlled Senate. The Senate Foreign Relations committee voted against the resolution in October. Consequently, the Senate never voted on the freeze as a separate piece of legislation. One attempt was made to add it to a debt ceiling bill as an amendment; on 31 October the effort was defeated by a vote of 40–58.

Legislating Strategic Programs: The MX

Passage of the freeze resolution was of course only one goal of the peace movement. Almost every peace and arms control group also sought to halt the development, production, and deployment of several individual nuclear weapons programs. Although the official freeze organization had originally sought to avoid the appearance of unilateralism, the freeze mobilization now provided the backdrop and framework for a frontal assault on specific elements of the Reagan nuclear weapons program. Heading the list of targets was the MX missile, with the B-1 bomber, the Trident submarine, cruise missiles, and the Euro-missiles following close behind. The most significant challenge to Reagan's buildup was the campaign to stop the MX. Although its extensive use of congressional education, lobbying, and public opinion warfare came close to derailing the program, congressional affinity for procedural solutions to controversial problems saved the beseiged missile.[22] The politics of the MX exemplify a pattern of defense policy decisions in which congressional criticism of a questionable Pentagon program leads to a congressionally initiated compromise, which often saves the controversial program and hastens the advent of a follow-on program.

The MX was jeopardized by its convoluted history of deployment plans and justifications.[23] The conservatives' argument about ICBM vulnerability was becoming influential while the MX was in its early phases of research and development in the 1970s. This argument

22. Les Aspin discusses critically the congressional penchant for procedural solutions in his "The Defense Budget and Foreign Policy," *Daedalus* 104 (Summer 1975): 155–74. Ironically, Aspin himself, as we shall see, has been the architect of procedural compromises on nuclear arms in recent years. In particular, his efforts had much to do with the eventual salvation of the MX.

23. On the origins of MX and its problems through the Carter administration see Herbert Scoville, *MX* (Cambridge, Mass., 1981); also John Edwards, *Superweapon* (New York, 1982). On the MX since 1980, with emphasis on presidential-congressional relations see Lauren H. Holland and Robert A. Hoover, *The MX Decision* (Boulder, Colo., 1985).

held that a new way of basing or defending ICBMs to protect them from a Soviet first strike was essential to any future land-based missile program. That position in turn made the MX missile less important than the mode of deployment or the basing scheme. The MX could not be a simple silo-based follow-on to the Minuteman ICBM. Carter responded with the multiple protective shelters proposal, the gargantuan shell game to be built and played in the deserts of Nevada and Utah. Many among both conservatives and liberals opposed Carter's plan—liberals because it was a new offensive weapon guaranteed to attract media attention and stir controversy, conservatives because a Democratic administration had proposed it and because it was so outrageous that they could easily call for a less complicated, if not less expensive, program. Political pressure from back home was also forcing some western politicians to oppose the multiple shelters system, which would have consumed a vast area of land in both Utah and Nevada and used enormous quantities of the region's most precious resource, water, for its construction. The anti-MX campaign in the West was as much an environmental as a peace issue; ranchers and farmers joined with pacifists and environmentalists. Even two of Reagan's friends and closest allies, senators Jake Garn of Utah and Paul Laxalt of Nevada, lobbied against the deployment plan.

Amid increasing critical publicity and compelled to find an alternative to the deployment plan he had criticized as a candidate, President Reagan resorted to his first of many presidential commissions. During 1981, Defense Secretary Weinberger and President Reagan formed the Townes Commission, a bipartisan panel of mostly technical experts headed by Nobel laureate physicist Charles Townes. Its job was to study various basing options to replace the multiple protective shelter plan. The panel came up with no single best option. Instead it recommended the study of several alternatives, including airborne deployment, deep underground basing, ABM point defense, and closely spaced basing. Congress specifically withheld money for the development of airborne deployment when that option appeared to be the administration's favorite. Reagan opted to put a few dozen of the missiles in hardened existing silos as an interim solution, pending a decision on a permanent basing mode. Congress continued to erect funding roadblocks in 1981 and 1982, conveying the message that little money would be spent on the MX until the administration had produced an acceptable basing scheme.

In November 1982, after considering some thirty basing options, the administration unveiled its plan for closely spaced basing, or Dense Pack, in which one hundred of the missiles would be packed as tightly as possible, rather than spread as widely as feasible. This apparent reversal of the whole logic behind the MX's original justification only confused the attentive public and added to congressional opposition. Quickly labeled "dunce pack" because of its dubious technical justifications, the scheme never had a chance. In December, almost exactly a year after it had stymied the president's interim plan, Congress, most significantly the Republican Senate, rejected the first long-term plan. This time, it insisted that it alone would be the judge of the new plan, which it asked the executive to submit by March of 1983.[24]

Meanwhile, public interest groups and local coalitions had mounted an extensive campaign against the weapon. This effort eclipsed the ABM movement of the late 1960s and the B-1 campaign of the mid-1970s. The MX—because of its theoretical first-strike capability, eight to ten warheads, extraordinary price tag, and close association with sanguine visions of nuclear combat—became synonymous with the nuclear threat. The media assailed the basing imbroglio. Few politicians could defend the program; many treated it with open sarcasm. Powerful as this opposition was, it could not kill the MX.

Instead, Congress and the president saved it with a procedural solution embodied in the rejection of Dense Pack and the request for another plan. Formed after Congress's refusal to fund the MX missile in December 1982, the Scowcroft commission (formally the President's Commission on Strategic Forces) was to "review the purpose, character, size, and composition of the strategic forces of the United States." The commission was composed of prominent former State, Pentagon, and CIA officials from both Republican and Democratic administrations.[25] Unlike the Townes Commission, which had failed

---

24. The House voted 245–176 to delete $998 million in procurement funds for the first five missiles. The Senate voted 56–42 to retain funding but to withhold it until Congress approved a new basing plan by joint resolution. The conference combined the major provisions of each, while allowing expenditure of R&D money on the missile itself.

25. Brent Scowcroft, the chair, was President Ford's national security advisor. Commissioners included Alexander Haig (Reagan's first secretary of state), William Perry (undersecretary of defense under Carter), and Richard Helms (Nixon's CIA director). Among the senior counselors were Harold Brown (Carter's defense secretary), Lloyd Cutler (Carter's presidential counsel), Henry Kissinger, and James Schlesinger. The notables of the commission met twenty-eight times, with many smaller conferences, and heard from over two hundred technical experts.

in its attempt to manufacture a technical solution, the Scowcroft panel had a more explicitly political mission: to produce a bipartisan compromise on strategic policy that would save the MX program. In April the panel submitted its report, which centered on three recommendations.[26] The first was that the hardened, fixed-silo approach be used for one hundred MX missiles as an interim solution for ICBM modernization. The second recommendation was that the United States develop a small, mobile, single-warhead missile, the Midgetman, to be the long-term solution to ICBM modernization and vulnerability. Third, the commission advised that future arms control negotiations concentrate on the reduction of warheads to enhance stability.

The panel's justifications for immediate deployment had more to do with the political psychology than the physics of nuclear competition. Commissioner Scowcroft and Counselor Harold Brown admitted in Senate hearings that, had purely military considerations guided their judgments, they would have opted for large, mobile missiles; but political factors compelled an alternative solution.[27] One of the commission's principal justifications for the immediate MX deployment was its utility as bargaining leverage in Geneva. Another was that America could ill afford to demonstrate irresolution on matters of national security. Finally, to justify the interim basing plan, these members of the cold war elite had to amend an argument they and their colleagues had fashioned a few years earlier. The panel played down previous claims about the Soviet ICBM threat. Most important in this effort was the commission's revised estimate of vulnerability. Although the media and enemies of the missile exaggerated the extent to which the commission disowned the window-of-vulnerability argument, the report does mitigate the threat. The report states that the Soviets probably possess the capability to destroy all U.S. ICBMs but that this vulnerability is "not a sufficiently dominant part of the overall problem of ICBM modernization to warrant other immediate steps."[28]

Moderate and conservative Democrats, including Les Aspin and Sam Nunn, fashioned the procedural compromises that led to the

26. President's Commission on Strategic Forces, Brent Scowcroft, chair, *Report of the President's Commission on Strategic Forces*, 6 April 1983.
27. U.S. Senate, Committee on Armed Services, *MX Missile Basing System and Related Issues*, 98th Cong., 1st sess., 18, 20, 21, 22, 26 April, 3 May 1983, pp. 8, 19, 20; see also Steve V. Roberts, "Leader of MX Panel Says Politics Played Key Role in Decision," *New York Times*, 19 April 1983, pp. 1 and D 23.
28. President's Commission, *Report of the President's Commission*, p. 17.

Scowcroft solution and the eventual congressional approval of the MX.[29] In the House, Les Aspin, respected as one of the most knowledgeable liberal critics of the Pentagon, led a small group of Democratic representatives in support of the small, single-warhead missile.[30] Aspin met with James Woolsey, a member of the commission and a navy undersecretary under Carter, and Scowcroft several times during the panel's tenure. Aspin mustered support for the idea on the Hill well before the commission finished its report.

In the Senate, under the sponsorship of Nunn, William Cohen, and Charles Percy, the build-down approach to arms control had gained much support. The build-down, which had originated as an alternative to the freeze, called for the destruction of two deployed warheads for every new warhead deployed (with the ratio of two-to-one subject to variation depending on the weapon). The ostensible purpose was to allow modernization and to move away from mutual land-based vulnerability while achieving real reductions. Build-down meshed nicely with Scowcroft's recommendation that future arms control negotiations be based on reductions in warheads.

The Aspin group, in particular, sold the vote on the MX as a package deal that included presidential commitment to Midgetman and the new approach to arms control, implicitly to build-down—in other words, the whole set of Scowcroft recommendations. To assure their colleagues, both the Senate and House groups sent letters to the White House asking the president for his commitment to the Scowcroft package, not simply MX deployment. Reagan sent replies to both groups endorsing all three recommendations but committing himself to nothing.[31] The support of the moderate Democrats and their insistence that Reagan would follow through on the complete package contributed to the comfortable margin in the Senate (59–39) and probably produced the winning margin (239–186) in the House on the 24 May votes releasing the MX funds. The administration's

---

29. An excellent chronicle of the politics behind the MX decisions in 1982 and 1983 with a focus on the Scowcroft commission and Aspin's maneuvering is Elizabeth Drew's "A Political Journal," *New Yorker*, 20 June 1983, pp. 39–75; see also "MX Pulls Through Turbulent Year in Congress," *Congressional Quarterly Almanac, 1983*, pp. 195–205.

30. In addition to Aspin, the "gang of six" included Thomas Foley, Albert Gore, Dan Glickman, Vic Fazio, and Richard Gephardt. Gore had been promoting de-MIRV-ing and the small, single-warhead missile since 1982; see his testimony in U.S. Senate, Committee on Foreign Relations, *Nuclear Arms Reduction Proposals*, 97th Cong., 2d sess., 29 and 30 April, 11–13 May 1982, pp. 214–38.

31. Reagan's 11 May letter to Representative Norman Dicks is reprinted in *Congressional Quarterly Almanac, 1983*, p. 32-E.

compromises cost it nothing, and the Scowcroft maneuver foiled the peace movement's campaign against the missile just when victory was at hand.

Moreover, the timing of the votes on MX and the freeze was a deliberate construction. The House leadership made sure that the MX vote followed the vote on the freeze. That way, once many members had expressed largely symbolic support for arms control, they could quickly turn around and vote for the MX in an attempt to balance the ambiguities of public opinion. Both the freeze and the MX followed the struggle in April over the confirmation of Kenneth Adelman as director of the Arms Control and Disarmament Agency. The Senate hearings on Adelman had become an important element in the public debate on the sincerity and competence of Reagan's arms control efforts. Despite the controversy, the Senate had confirmed Adelman on 14 April by a vote of 57–42. So the ambiguous House freeze victory of early May was bracketed by two defeats.

The MX program continued to sputter forward, albeit in a reduced form, and neither build-down nor START ever bore fruit under Reagan. As chair of the House Armed Services committee, Aspin remained at the center of the ongoing MX funding-procurement debate by his insistence that further procurement of MXs be tied to accelerated development of the Midgetman.[32] Aspin and other members of Congress felt compelled to include such prescriptions because the air force and the Reagan administration did not support the Midgetman program; they wanted the MX first and foremost. The irony of members of Congress, and especially a core of Democrats, having to coax the administration into funding a second multi-billion-dollar ICBM program led some to dub the small missile "the Congressman." The MX-Midgetman debate produced some odd divisions and alliances in Congress, with some liberals supporting a limited MX to avoid Midgetman, while others backed Midgetman R&D funding because it was less expensive than MX deployment. Conservatives often were divided over the best way to get both missiles. By the beginning of the Bush administration, both missile programs were being developed for mobile basing, with the MX to be deployed on trains and the Midgetman on large trucks.

This example illustrates a pattern that emerged more than once in the 1980s. As with the MX, opposition to the B-1 bomber by the

---

32. Aspin said that Midgetman is "our priority strategic weapon in development." Paul Mann, "House Armed Services Committee Adopts 5% Real Decline in Defense Spending," *Aviation Week and Space Technology*, 7 July 1986, p. 24.

peace movement lobby led eventually not to its termination but to an agreement that its deployment be contingent on the development of a new and better platform. The B-1's questionable advantages over the B-52 produced support for an even more advanced aircraft, thus accelerating development of the Stealth bomber before deployment of the one hundred B-1s was complete.[33] Instead of bringing about the termination of the original program, the movement's critique and attack only hastened the development of a follow-on program.

Just as the vote on the House freeze resolution marked the high point of the peace movement's efforts to legislate arms control policy, the vote on the MX foretold the future of legislative efforts to derail nuclear weapons programs. By the end of 1983 it was clear that, while the freeze had lost momentum in the legislative process, the campaign had at least compelled Reagan to take symbolic action on arms control. He had introduced and endorsed an unprecedented series of arms control proposals and related initiatives in the course of about a year. While achieving little abroad and in Moscow, the zero option, START, Star Wars, and build-down had a significant impact at home. These initiatives were partly motivated by and directed at the most significant threat to Reagan's foreign policy agenda: the freeze. Reagan's proliferation of proposals cost the administration little, and after two years, the peace movement had nothing concrete to show for its efforts besides a more informed and concerned public. But the movement had caused concern in the White House and in Republican circles everywhere. In fact, it had helped to produce the most distinctive policy decision of the Reagan presidency, the commitment to Star Wars. But more important, it had put the administration on the defensive on arms control policy and had considerably dampened public enthusiasm for further military buildup.

Arms Control and Electoral Politics

Toward the 1984 Elections

At its 1983 annual convention, the freeze movement, buoyed by success in its public education campaign but frustrated by a reluc-

---

33. For histories of the B-1 and B-2 bombers and the relationship between them see Nick Kotz, *Wild Blue Yonder: Money, Politics, and the B1 Bomber* (New York, 1988); Frank Greve, "The Stealth Juggernaut," *San Jose Mercury*, 12 March 1989.

tant Congress, decided to expand its political horizons. Because the freeze campaign had begun as a public education effort, the leadership of the movement had been reluctant to try involvement in electoral politics. At the 1983 convention, however, they decided to make explicit what would have been implicit in their election-year activities anyway. They formed a parallel organization, a PAC called Freeze Voter 1984. In addition to raising direct campaign contributions, Freeze Voter was to canvas the public, extolling the freeze, informing citizens of the candidates' positions on it, and urging them to register to vote. The goal was to "elect in 1984 a President and Congress who will actively support the freeze."[34] However ostensibly bipartisan, the freeze movement looked forward to 1984 and the presidential and congressional elections with an awareness that its hopes rested on whatever influence it could muster within the Democratic party. This reality applied as well to the New Politics coalition as a whole, of which the peace movement was the critical constituent. Together, their goals were threefold: First, the peace movement and the New Politics coalition sought to capture control of the Democratic party's foreign and military policy platform, thereby making the freeze the litmus test for any presidential hopeful. Second, they sought to change the composition of the Congress enough to stem the buildup and adopt the freeze as a binding resolution, whether a profreeze candidate won the presidency or not. Third, the New Politics coalition and peace movement would demonstrate their collective importance to the party by electing the resulting profreeze candidate to the presidency.

Fundamental to the first goal was a change in the movement's strategy. For the first two years, the freeze coalition of peace groups had built its mostly white and middle-class base and drawn upon other New Politics movements for support. As the arms race became the dominant issue, other New Politics groups, especially environmentalists, had jumped on the bandwagon. Going into the election season, however, the peace movement sought to broaden its mass base through coalition building, as in the 1983 endorsements of Jobs with Peace Week and the Martin Luther King March on Washington. They adopted a resolution criticizing Reaganomics[35] and reached out to blacks, Jesse Jackson, and his nascent Rainbow Coalition. In stressing budgetary priorities, this potential alliance found common ground in mutual opposition to increased military spend-

34. Conference declaration quoted in *Nuclear Times*, March 1983, p. 7.
35. David Corn, "Freeze Sets Daring Strategy," *Nuclear Times*, March 1983, pp. 6–9.

ing at the expense of social programs. The endorsement of Jobs with Peace was the beginning of an attempt to forge links with organized labor in basic industries feeling the pinch of international competition and not experiencing any benefits from the arms buildup.

An action by the Catholic bishops provided an opening for the movement to gain support in the Catholic population. In their pastoral letter from May of 1983, the bishops endorsed almost every key peace movement proposal, including the freeze, opposition to first-strike weapons, no-first-use, and a comprehensive test ban.[36] The alliance with Catholics provided new resources, publicity, and respectability. Moreover, because Catholics' traditional support for the Democratic party had been eroding, Catholic support for the peace movement showed how important the New Politics could be in rebuilding the Democratic coalition. Some Protestant groups with historical commitments to peace and disarmament, for example, the Quakers and Unitarians, were natural allies; and other Protestant churches, among them Episcopalians and Methodists, soon endorsed the movement's goals. These denominations added to the movement's influence among white middle- and upper-class liberals, especially in the Northeast, a critical area of Democratic strength.

Some women's groups had been among the early leaders of the new peace movement, like the Women's International League for Peace and Freedom and Women's Action for Nuclear Disarmament. Through coalition building, more mainstream feminist organizations joined with the peace movement to attack the military budget and its deleterious effects on women, especially by relating military spending to the feminization of poverty. NOW endorsed many of the peace movement's proposals. Political commentators had ascribed the gender gap in voting largely to women's greater concern over Reagan's hawkish policies. Because it appeared to be a major threat to the GOP's electoral fortunes in 1984, the alliance of women's groups with the peace movement added to the power of both.

Thus alliances were formed with other groups affected by the clash between the welfare and warfare states. In this way, the freeze was not just an attempt to alter U.S. strategic defense and arms control policies; more significant, it quickly became the political fulcrum with which the New Politics coalition, as the core of the liberal wing

---

36. National Conference of Bishops, *The Challenge of Peace*, 3 May 1983.

of the Democratic party, would reassert its power. The freeze reignited and united the New Politics and fostered alliances among all those adverse to and adversely affected by the Reagan revolution. The freeze, as the symbolic shorthand for all issues of war and peace, became the political rallying point for the New Politics coalition and its allies to assert control over the Democratic party and to stem the rising tide of conservatism manifested by Reagan and his political agenda.

Although the peace movement and the freeze were having a limited impact on defense and arms control policies, their influence on national politics had not yet peaked and was still taking shape. Regardless of what had happened in Congress, the impact of the freeze and the arms control issue on the 1982 elections affected the strategy of politicians thinking about 1984.[37] Consequently, while the freeze movement was lobbying Congress and mobilizing public opinion, the Democratic party, as represented by its leadership and party conventions, underwent a sea change in its military policy agenda between 1980 and 1984, a transformation that reflected the growing power of the peace movement and its issues. At the 1980 convention, Representative Dante Fascell had opened his remarks in support of the platform's foreign policy plank by proudly noting that the Democrats had reversed the post-Vietnam decline in defense spending. Even more illustrative of the leadership's relatively sanguine attitude toward cold war policies had been the defeats of minority reports on the convention floor. The first was Minority Report 20, which opposed development of the MX missile; it lost by a considerable margin on a roll call. The second plank, Minority Report 21, never made it to a roll call. This plank was essentially the bilateral freeze proposal, which had been gathering support among the leadership of the nascent peace movement.[38] The plank had no sponsors among the Democratic leadership. John Kenneth Galbraith and retired rear admiral Gene LaRocque, both elite critics of nuclear weapons and policy, and soon to be spokesmen for the nuclear

---

37. As Chip Reynolds, director of Freeze Voter '84, said in an interview, the movement and the issue perhaps "had the illusion of having greater organization and impact than may have been true" based on its success in 1982, which produced a "cache" of influence going into 1984.

38. Minority Report 21 read: "The Democratic Administration will pursue an immediate freeze applying to all nuclear weapons states, on all further testing and deployment of nuclear weapons and delivery systems within the limits of verifiability." *Official Report of the Proceedings of the Democratic National Convention*, New York, 11–14 August 1980, p. 444.

peace movement, spoke for the plank. Governor Charles Robb of Virginia, soon to be a leader of conservative Democrats, castigated it. The plank failed on a voice vote.

By summer 1982, things had changed. At the Philadelphia Democratic National Conference in June, the national security drafting committee had endorsed the freeze proposal. From then to the beginning of the 1984 election season, support for the freeze grew. In a September 1983 speech, Charles Manatt, the chair of the Democratic party, had stated that unlike the Republicans, "the Democratic Party calls for a mutual, verifiable freeze . . . now."[39] Every candidate, except Reuben Askew, had approved of Manatt's announcement before his speech.

During the primary campaigns, all of the Democratic contenders made, or were compelled to make, arms control a focal issue. Alan Cranston and George McGovern based their campaigns on the power of nuclear war issues and the movement attached to them. Jesse Jackson aligned the Rainbow Coalition with the peace movement by endorsing the freeze and especially by calling for real reductions in defense spending. Although most among the traditional Democratic leadership were eager to link arms with the movement, some were less keen to embrace the actual freeze proposal. During their short-lived campaigns, Reuben Askew, John Glenn, and Ernest Hollings spoke in generalities about the dangers of nuclear war and the need for arms control. Askew was the only hopeful who at no time endorsed the freeze. The front-runners, Hart and Mondale, had both advocated a freeze but often preferred to concentrate on other ideas and themes. Hart favored negotiations on nuclear war prevention and talked about military reform. Mondale attacked Star Wars and the failure of Reagan's arms control efforts and called for an antisatellite weapons ban and a war on Pentagon waste. The future nominee, albeit less publicly, also advocated real increases in defense spending and development of the Stealth bomber and the Midgetman missile. Nevertheless, there was even some bickering among Jackson, Hart, and Mondale over who was first to endorse a freeze.

As arms control lobbyist Christopher Paine noted, "The freeze movement . . . reached its peak of organization, expenditure, and political influence during the 1984 election, long after the media had

---

39. Dan Balz, "Party Chairman Puts Democrats behind a Freeze," *Washington Post*, 21 September 1983, p. 21.

lost interest in it."[40] Indeed the movement was still growing in membership and financial resources, which translated into electoral influence. During the 1984 election cycle, the national Freeze Voter PAC and its local affiliates in forty states combined to raise $3.5 million, making it one of the largest independent PACs. This money went to nearly three hundred grass-roots organizations at work in 244 House and 20 Senate races, involving more than twenty-five thousand volunteers.[41] Other peace groups, including SANE and Council for a Livable World, contributed well over $2 million to congressional campaigns, the lion's share to Democrats.

Not surprisingly, the peace movement's influence was quite evident at the Democratic convention. Alan Cranston, the favorite son of much of the mainstream peace movement, helped organize a committee of peace movement intellectuals and leaders to draft a plank on nuclear arms control for the Democratic platform.[42] The plank turned the complete agenda of the freeze movement into a series of fourteen succinct steps for the future Democratic president to implement, centering around a series of bilateral moratoriums, or "quick freezes," to be followed by a comprehensive freeze and other measures. The Peace Roundtable, a group of about forty peace, arms control, and other liberal organizations, made a similar effort and reached consensus on a ten-point program, which it submitted both to the Democratic Platform committee and to the Republican National Committee as proposed elements for a "Peace Plank."

The Democratic convention embraced virtually the whole arms control agenda of the peace movement. The moratoriums and comprehensive freeze are foremost in the "Peace, Security, and Freedom" plank. In fact, the platform specifically directed the new Democratic president to initiate the moratoriums on 20 January 1985 as one of his first acts in office. The platform also includes planks opposing the MX and B-1 bomber, antisatellite weapons, and Star Wars. Military reform ideas, clearly second in priority, dominate the

---

40. Christopher Paine, "Lobbying for Arms Control," *Bulletin of the Atomic Scientists*, August 1985, pp. 126–27.
41. Mark Hertsgaard, "What Became of the Freeze?" *Mother Jones*, June 1985, p. 46.; Paine, "Lobbying for Arms Control," p. 126.
42. Council for a Livable World, *Draft Plank for Democratic Party Platform on Nuclear Arms Control* (Boston, 1984). The drafting committee was Jerome Weisner (former president of M.I.T. and board member of Council for a Livable World), Randall Forsberg, Jerome Grossman (president of the council), and Ruth Adams (editor of the *Bulletin of Atomic Scientists*), with special advisors McGeorge Bundy, George Kennan, and Carl Sagan.

section on defense policy. The platform directly addresses the character and power of the New Politics movement: in the same section that calls for the freeze, a passage states that the Democratic president will "strengthen broad-based, long-term public support for arms control by working closely with leaders of grass-roots, civic, women's, labor, business, religious and professional groups, including physicians, scientists, lawyers, and educators."[43] Nevertheless, many in the party leadership sought to avoid the charges of weakness leveled against their party by the Republicans. This faction was able to defeat attempts to put the party on record for real decreases in defense spending instead of a slowdown in real increases. The mainstream of the Democratic party wanted to attack the Reagan arms control record but still appear strong on defense.[44] In this, the party mirrored the divisions in public opinion, with its majorities in favor of both a nuclear freeze and various measures of a strong if not superior military capability.

The Reagan Record and Democratic Strategy

The Reagan campaign team chose the cautious strategy, to run on the president's record rather than articulate a vision for a second revolution.[45] With the exception of Reagan's, perhaps purposefully vague "You ain't seen nothing yet," the campaign stuck to the accomplishments of the administration. "It's Morning in America" evoked an image of the last four years restoring pride, strength, and economic prosperity and left unsaid what sort of day was to follow morning. Instead, the campaign focused on "Leadership That's Working." In an echo of Reagan's 1980 question "Are you better off now than you were four years ago?" the 1984 theme was an assertive "You *are* better off now than you were four years ago." The

---

43. "Text of the 1984 Democratic Party Platform," *Congressional Quarterly Weekly Report*, 21 July 1984, pp. 1747–80.

44. See the debate and vote on Minority Report 4 and remarks by former Senator James Cavanaugh: *Official Proceedings of the 1984 Democratic National Convention*, pp. 256–60. Interviews with the chief Washington lobbyists of two arms control groups, John Isaacs of the Council for a Livable World and Chip Reynolds of Freeze Voter 1984, reenforced the impression that most of the Democratic candidates, and the party leadership, were ambivalent about the freeze movement. The Democrats saw it as one of the most salient issues and wanted to tap into the power and popularity of the movement without being captured by it or endorsing its complete agenda.

45. John H. Aldrich and Thomas Weko, "The Presidency and the Election Process: Campaign Strategy, Voting, and Governance," in Michael Nelson, ed., *The Presidency and the Political System* (Washington, D.C., 1988), pp. 256–59.

Reagan team chose to run on Reagan's popularity, his image as a strong leader, and the state of the economy.

Mondale ran into considerable difficulties when he tried to use the economy and even the deficit as issues, portraying the economic recovery as partial and transitory. The attempt to translate complicated arguments about a "Swiss cheese" economy and "hourglass" wage structures into gripping political slogans failed next to Reagan's quick and easy references to overall employment and inflation.[46] The Democrats argued that economic growth would ultimately come to a grinding halt and plummet from the burden of the deficits, but the underlying arguments about debt, interest rates, investment, and exports were difficult to get across. Warnings about dire consequences at some future date, with no contemporary signs of pending doom, simply did not hit home, even with analogies to the family checkbook. Despite Mondale's promise of quick and decisive action to lower the deficit, a majority in public opinion polls thought Reagan would do a better job of balancing the budget.[47] Mondale had already crossed the political Rubicon, however, when he decided to be honest with the American public, in his acceptance speech at the convention, about what was required to solve the deficit problems: "Mr Reagan will raise taxes, and so will I. He won't tell you. I just did."[48] Many saw this as one of Mondale's worst mistakes, and the Republicans quickly went on the attack.[49]

The Mondale campaign searched for themes that would work and issues on which Reagan was vulnerable. In essence what they came up with were two issue umbrellas: "fairness" and "war and peace."[50] Fairness encompassed the deficit and, more directly, Reagan's cuts in programs and his tax breaks for corporations and the rich, with a dash of other issues such as the environment and Reagan's pandering to corporate polluters. The second umbrella issue often ranked

---

46. For example, in his acceptance speech at the convention Mondale said that while the rich are better off, working Americans are worse off, "and the middle-class is standing on a trap door." Obviously, not many middle-class Americans could see the door. *1984 Democratic National Convention*, p. 489.

47. Paul J. Quirk, in his "The Economy: Economists, Electoral Politics, and Reagan Economics," in Michael Nelson, ed., *The Elections of 1984* (Washington, D.C., 1985), p. 177, offers further analysis of how Reagan benefited on economic issues.

48. *1984 Democratic National Convention*, p. 489.

49. Paul R. Abramson, John H. Aldrich, and David W. Rohde, *Change and Continuity in the 1984 Elections* (Washington, D.C., 1985), pp. 55–56.

50. On the Mondale strategy see Bernard Weinraub, "Mondale Plans to Focus on Issues Where He Says Reagan Is Weak," *New York Times*, 9 September 1984, pp. 1, 35.

close to the economy in its importance to potential voters throughout the campaign. In fact, in one Gallup postelection survey, more voters (25 percent) picked nuclear weapons and the arms race (out of twelve choices) as the most important national problem over any other single issue.[51] The Mondale forces formulated three major arms control themes for the campaign. First, Mondale hammered away at the Reagan administration's failure to negotiate any arms control treaty with the Soviet Union, unlike all previous administrations. Furthermore, Reagan was the first postwar president not even to meet with his Soviet counterpart. Second, the Democrat offered the freeze as the Democratic arms control solution. Third, Mondale repeatedly attacked Reagan's Star Wars program. In a September interview, Mondale said that he saw "arms control as the central issue in this campaign."[52] He announced that on his first day in office he would "call on the Soviet leadership to meet with me within six months for . . . negotiations to freeze the arms race and to begin cutting back the stockpiles of nuclear weapons."[53] In this way, Walter Mondale's campaign attempted to convert Reagan's pointed inquiry of 1980, "Are you better off than you were four years ago?" into "Are you living in a safer world than you were four years ago?"[54]

In the first debate on domestic policy, Mondale used his concluding remarks to contrast Reagan's are you better off theme against the harsher realities Mondale saw, beginning with "Are you better off with this nuclear arms race? Will you be better off if we start this star wars escalation into the heavens?"[55] In the second debate, while sounding fairly hawkish on Central America, Mondale again tried to

51. Aldrich and Weko, "Presidency and the Election Process," p. 259.
52. Bernard Weinraub, "Mondale Reports He and Gromyko Are to Meet Soon," *New York Times*, 17 September 1984, p. B12. Mondale's pollster, Peter Hart, said the war and peace issue "stands out as an absolutely fundamental issue where the voters see Reagan as off in the wrong direction and Mondale headed in the right one"; Weinraub, "Mondale Plans," p. 35.
53. Bernard Weinraub, "Mondale Pledges Immediate Effort for Arms Freeze," *New York Times*, 6 September 1984, p. 1.
54. In a 17 September speech, Mondale said, "The fact of it is that four years of Ronald Reagan has made this world more dangerous. Four more years will take us closer to the brink." Quoted in Christopher Madison, "Mondale Asking Voters If They Feel Safer Now Than They Did Four Years Ago," *National Journal*, 20 October 1984, p. 1960. During his acceptance speech, Mondale appealed specifically to those who had voted for Reagan in 1980: "You did not vote for an arms race. You did not vote to turn the heavens into a battleground"; *1984 Democratic Convention*, p. 488.
55. Transcript of Louisville presidential debate, *New York Times*, 8 October 1984, p. B7.

portray the president as a misinformed commander in chief who had failed to negotiate with the Soviets and whose "definition of national strength is to throw money at the Defense Department."[56] Mondale called for a nuclear freeze, in contrast to Star Wars, and for renewed emphasis on conventional forces. Both Star Wars and the freeze were major topics in the debate.

War and peace and arms control became by default the issues with the strongest potential for Mondale and his campaign. As a Mondale advisor put it, "I don't know whether foreign policy is a winner or a loser for Mondale, but right now, he hasn't got much choice."[57] As November approached, much of the party's hope rode on the power of the New Politics coalition, in its incarnation as the peace movement. It alone seemed to have the power to link elements of the old Democratic coalition, including blacks, laborers, and Catholics, with other potential voters within the ranks of women, "yuppies," and moderate Republicans.

While the Republicans adhered largely to the military policies and platforms that had brought success in 1980, the Democrats altered course to port in an effort to make headway against the Republican coalition. Mondale sought to exploit the one set of issues on which he appeared to hold the advantage. The election revealed, however, that Mondale could not overcome other Reagan advantages and that Reagan's vulnerability on defense and arms control was exaggerated. Reagan won on the evaluations of his job performance and effectiveness, with the economy perhaps the principal issue working to his advantage. Peter Hart found that while 70 percent of Reagan voters approved of his economic policies, 41 percent disagreed with his plans for further defense budget increases, and 40 percent disagreed with his approach to arms control.[58] A postelection survey showed that Reagan beat Mondale nonetheless, by 48 to 37 percent, on which candidate was better able to handle the issue of nuclear weapons and the arms race.[59] Moreover, in one preelection survey,

---

56. Transcript of Kansas City presidential debate, *New York Times*, 22 October 1984, p. B5. Mondale jumped on Reagan's claim that submarine nuclear missiles were safer because they were recallable; Reagan used one whole rebuttal time to explain his remark about submarine missiles.

57. Leslie H. Gelb, "Mondale Shift to Right on Foreign Policy Seen," *New York Times*, 12 September 1984, p. B9.

58. "Moving Right Along? Campaign 1984s Lessons for 1988," *Public Opinion*, December/January 1985, p. 10. Analysts agreed the economy was the most influential issue and benefited Reagan by wide margins.

59. Aldrich and Weko, "Presidency and the Electoral Process," p. 262.

potential voters were asked which ticket, regardless of their actual favorite, would do the better job of keeping America militarily secure. The Reagan-Bush team overwhelmed the Mondale-Ferraro ticket by 57 to 16 percent. The Republican ticket also received a better rating when the issue was keeping peace in the world.[60] One election-day survey had 14 percent of all voters listing military spending as the one issue about which they most liked the stand of their candidate, a figure matched only by the 14 percent listing governmental spending. Of that 14 percent who chose military spending, 75 percent voted for Reagan and only 25 percent for Mondale.[61] Thus nearly 11 percent of Reagan's supporters ranked military spending (spending, not national defense) as the primary reason for voting for Reagan.

Despite the efforts of the peace movement and the popular support for the freeze, and despite increasing public reluctance to support a sustained military buildup at the price of unprecedented deficits, Reagan was still able to translate his national security policy into political support. The public's ambivalence on the issues of national security and arms control tended to work to the administration's advantage. Not long after the election, when asked if the problem of national defense had gotten better or worse as a result of Reagan's policies, 62 percent said it had gotten better, while 10 percent said worse and 20 percent said it was the same. His positive rating for defense nearly equaled that for inflation (64 percent) and exceeded his positive ratings for reducing government bureaucracy, balancing the budget, and solving unemployment and energy problems. Even 54 percent of Democrats thought he had improved national defense.[62]

Yet the freeze and the peace movement had been a force to be reckoned with at the congressional level. The campaign labor and money provided by the movement had an impact on many races. Five of the seven new senators pledged to support the freeze, including Paul Simon and Tom Harkin, who both credited Freeze Voter activism with helping them win. Freeze Voter candidates won in twenty-five of thirty-five targeted House races. But success at the

60. *Los Angeles Times* poll, 12–15 October 1985.
61. ABC News/*Washington Post* poll, reprinted in Scott Keeter, "Public Opinion in 1984," in *The Election of 1984*, ed. Gerald Pomper et al. (Chatham, N.J., 1985), p. 96.
62. This was an 8 percent increase in his support from when the same question was asked in 1982; see *Gallup Report* 235 (April 1985): 4–6.

congressional level was overshadowed by the landslide defeat of the Democratic attempt to recapture the presidency.

The Democrats Reconsider

After its drubbing at the presidential level, the Democratic party reevaluated its public appeal and the constituencies it had tried to unite during the election. Trying to exploit a gender gap had proven a hollow strategy.[63] The Rainbow Coalition had had a limited impact. And the peace movement and its central political initiative, the freeze, had failed to capture the imagination and votes of middle America. The party emerged ideologically confused on defense policy and hounded by a firm public perception of that ambivalence.

The advocates of a more centrist party gained ground and renewed the debate over the ideological future of the party. Playing a leading role in this effort was the Coalition for a Democratic Majority (CDM), headed in 1984 by Ben Wattenberg. At that time, its board was a list of luminaries in conservative Democratic politics: Scoop Jackson was the canonized dead; among the living were Nathan Glazer, Michael Novak, and Richard Pipes. The CDM had tried to outmaneuver the peace movement in 1984, especially its "peace plank," by drafting an alternative foreign policy platform for the convention. The document castigates the freeze, calls for 6–8 percent yearly real growth in military spending, and endorses every major nuclear weapons program (MX, Midgetman, B-1, Stealth, limited ballistic missile defense). Replete with the rhetoric of military reform, the platform calls for more efficient and stronger conventional forces as the best way to prevent nuclear conflict and deter war in general.[64] But CDM made little headway at that time.

After 1984, however, the rightward shift embodied by CDM's platform rapidly gained adherents in the party leadership. Two efforts to move the party in that direction got underway in early 1985: the Democratic Leadership Council and the Democratic Policy Com-

---

63. The author's research elsewhere shows that even before the 1984 elections it was apparent that the gender gap was the result of men defecting from the Democratic party at a faster rate than women; see Daniel Wirls, "Reinterpreting the Gender Gap," *Public Opinion Quarterly* 50 (Fall 1986): 316–30.

64. The document is entitled *Democratic Solidarity: A Draft 1984 Party Platform on Foreign Policy and National Defense* (N.p., n.d. [c. June 1984]). For an example from the publicity campaign the CDM launched in support of their platform see Leon Wieseltier, "The Pitfalls of the Peace Issue," *Washington Post*, 27 May 1984, p. C5.

mission. The Leadership Council, composed of 140 elected Democratic officeholders, sought to distill a program that would unite the party and free it from both New Deal liberalism and the New Politics agenda. Former governor of Virginia Charles Robb chaired the council; other prominent members included Bruce Babbitt (former governor of Arizona), senators Sam Nunn, Albert Gore, and Joseph Biden, and Representative Richard Gephardt (making the group a virtual roundtable of 1988 Democratic presidential hopefuls). In defense and economic policy the council's ideas reflected the themes Gary Hart raised in 1984 (industrial policy, competitiveness, and military reform), although Hart was not a member of the group.

Soon after he replaced Charles Manatt as chair of the National Democratic Committee, Paul Kirk initiated a reformulation of Democratic values and programs. The Democratic Policy Commission was created just for this purpose. Like the Leadership Council, the Policy Commission was composed exclusively of elected Democratic officeholders, one hundred of them from the federal, state, and local levels. Kirk readily admitted that he wanted elected officials instead of "representatives of single issues or limited agendas,"[65] meaning representatives of New Politics and its allies. Again the "neoliberal" ideas on defense and economics associated with Gary Hart, then the party's best prospect for 1988, pervade the commission's final report, "New Choices in a Changing America." The sections on defense, arms control, and foreign policy move to the right of Hart, however.[66] For example, the report makes only a veiled reference to Nicaragua (virtually ignoring the most controversial aspect of Reaganite foreign policy) and leaves the door open for military assistance to the Contras. While favoring adherence to SALT II and the ABM treaty, the report endorses the development of Midgetman

---

65. Quoted in Robin Tower, "Democrats Bring Out New Model," *New York Times*, 24 September 1986, p. 24. The Policy Commission also reflected an emphasis on the South and West in its leadership; the steering committee included national and local representatives from Arkansas, Idaho, Texas, Montana, Utah, Washington, Massachusetts, Vermont, Wisconsin, and Ohio. The most prominent members of the Leadership Council also hailed from the South and West: Robb, Gore, Babbitt, Nunn, and Gephardt.

66. The defense and foreign policy sections reportedly drew heavily on groundwork done by a committee headed by Les Aspin, who had recently voted for aid to the Nicaraguan Contras and engineered the compromise that saved the MX and gave birth to Midgetman. The final report's emphasis on Midgetman and hedging on Nicaragua indicate Aspin's influence. On Aspin's work before the final report see Phil Gailey, "Democratic Panel Shifts Foreign Policy Emphasis," *New York Times*, 3 July 1986, p. 21.

and the Stealth bomber, SDI research, and the continued deployment of Trident D-5 SLBMs. There is no mention of a freeze. Arms control comes up only in ancillary calls for a comprehensive test ban and negotiations to ban mobile missiles. The principal criticism of Reagan's defense policy is that it constitutes a "spending spree with no strategic rationale and no overall defense plan." The report does not question the need for Reagan's buildup, only the way in which the money was being spent. Beginning with the story about the officer on Grenada who had to use a credit card call to get naval fire support, the section on defense policy reiterates the standard arguments of the military reform movement: the importance of readiness, training, and organizational and procurement reform. It even reprints an abstract from Gary Hart's book on military reform, *America Can Win*. Even with Hart removed from the party's future only months after the report's appearance, the importance of the commission's work was apparent. Several of the potential Democratic presidential candidates for 1988 aligned themselves with this shift to the right.

These efforts by the moderate and conservative Democratic leadership were part of a broader movement, especially by members of Congress, away from the strident and polarized debate between the freeze movement and the advocates of the Reagan strategic modernization program. As the power of the freeze waned, the language of military reform waxed eloquent and powerful and began to displace the rhetoric of the nuclear nightmare, all the while attracting support from the media and the freeze movement's middle-class adherents. Many in the peace movement endorsed much of the reformers' agenda, especially their attacks on waste and fraud in weapons contracting, and in so doing, began to lose control of the national security debate. Whereas the movement had actually provoked national discussion about the ends of military policy (whether nuclear arms and deterrence can in any way add to our security or only jeopardize it), military reformers concentrated on the means of defense, and only the conventional means, to whatever end, be it deterrence or victory. The peace movement failed to produce a holistic position on national security. When the discussion began to turn from nuclear weapons and strategy to conventional arms and doctrine, the movement had no coherent response.

Military reform notwithstanding, a more immediate threat to the peace movement's agenda had emerged already: Star Wars, or the Strategic Defense Initiative. With a fresh landslide providing another endorsement of his defense policies, Reagan did not let the arms race stand still for the political opposition. His "vision" in 1983 of a world free of the threat of nuclear war, through strategic defense, was, from 1984 on, being translated into the single most important program in the Reagan military budget. Before 1983, the threats to the decade-long prohibition on defensive systems, as codified by the ABM treaty of 1972, were relatively remote. Star Wars was a real and present threat and as such went to the forefront of the peace movement's agenda. While the freeze campaign was still building its mass base and influence within the Democratic party, Reagan's announcement and promotion of Star Wars stole some of its public relations thunder and dissipated attacks on key offensive weapons, while simultaneously increasing the scope and impact of his military buildup. As the freeze proposal quickly faded after the 1984 elections, Star Wars gradually gained public approval and bureaucratic momentum. It is to the origins and development of the Strategic Defense Initiative that we now turn.

6

The Strategic Defense Initiative

On 23 March 1983, during a nationally televised address in defense of his military programs, the president announced his decision to "embark on a program to counter the awesome Soviet missile threat with measures that are defensive."[1] Reagan called "upon the scientific community in our country, those who gave us nuclear weapons, to turn their great talents now to the cause of mankind and world peace, to give us the means of rendering these nuclear weapons impotent and obsolete . . . I am directing a comprehensive and intensive effort to define a long-term research and development program to begin to achieve our ultimate goal of eliminating the threat posed by strategic nuclear missiles."[2] So began Reagan's most innovative initiative in military policy—Star Wars, as it was quickly dubbed by the press and its critics, or the Strategic Defense Initiative, as it was officially designated later that year. The ambitious five-year, twenty-six billion–dollar research program rapidly became the first order of business on the president's national security agenda and the center of national and international attention.

Most analysis of SDI explores and debates the strategic wisdom, technological feasibility, and international consequences of the pro-

---

1. Address reprinted in *Congressional Quarterly Weekly Report*, 26 March 1983, p. 632.
2. Ibid., p. 633.

gram.³ The origins of the initiative, as a case of defense policy formation, have received significantly less attention. Domestic political considerations, particularly the need to answer mounting opposition to the administration's defense and arms control policies, appear to have been the most important immediate consideration behind the decision. Furthermore, the evolution from policy initiation to program institutionalization reveals the extent to which the Star Wars case inverts the more common pattern of weapons program development. Although Star Wars began in 1983 as a hastily conceived insert in a presidential address, the Strategic Defense Initiative became one of the largest programs in the military budget before the end of the Reagan presidency.

The Origins of Star Wars

Two versions of the Star Wars decision emerged after the March 1983 speech and remain the dominant, and diametric, portrayals of the process. The first is essentially a rational actor model. The Reagan administration offered several strategic motivations for its decision to proceed down the path toward strategic defense. In the March 1983 speech, Reagan dwelt on the need to escape the perilous reality of MAD by constructing a world safe from nuclear holocaust. He also referred to the need to counter Soviet advantages in land-based offensive missiles. Other administration and Pentagon officials maintained that strategic defense would enhance deterrence. They also cited recent technological advances and the supposedly substantial Soviet work being done on strategic defense as compelling rationales for an accelerated research and development program. According to the administration, Star Wars emerged as a rational response to strategic realities and moral imperatives.

A different picture of the process emerged from press reports published soon after the March speech. These accounts harmonized around a few themes, producing the more popular view of the Star Wars decision, one that stressed the idiosyncratic elements: SDI was

3. For a sample of this vast and somewhat redundant literature see Steven E. Miller and Stephen Van Evera, eds., *The Star Wars Controversy* (Princeton, N.J., 1986); Sidney Drell et al., eds., *The Reagan Strategic Defense Initiative* (Cambridge, Mass., 1985); Robert Bowman, *Star Wars: A Defense Insider's Case against the Strategic Defense Initiative* (New York, 1986); Ivo Daalder, *The SDI Challenge to Europe* (Cambridge, Mass., 1987); Harold Brown, ed., *The Strategic Defense Initiative: Shield or Snare?* (Boulder, Colo., 1987); Franklin A. Long, Donald Hafner, and Jeffrey Boutwell, eds., *Weapons in Space* (New York, 1986).

the president's personal initiative, pursued outside bureaucratic channels and with little consultation of his principal advisors. Headlines such as "Aides Urged Reagan to Postpone Antimissile Ideas for More Study" and "President Overruled Advisors on Announcing Defense Plans" appeared.[4] According to these reports, even before the speech several administration officials and bureaucratic reports had publicly expressed pessimism about the technical feasibility and strategic wisdom of such a venture. Meanwhile, a few private individuals, in particular Edward Teller, had lobbied the already sympathetic commander in chief for a strategic defense program.[5] It was reported that Reagan personally drafted the Star Wars portion of the address, that only a few White House staff members reviewed it, and that top State and Pentagon officials were taken by surprise when they heard the speech.[6] The apparent importance of the president's vision and the conscious disregard of bureaucratic procedures and recommendations suggests decision making driven by personality and ideology—a nonrational, if not irrational, process.[7]

These two sharply contrasting descriptions are incomplete and exaggerated. The strategic rationality explanation overstates the influence of such motives; the personality/ideological explanation inflates the personal influence of Reagan and discounts the planning in the White House that preceded the decision. Moreover, the addition of other potential influences does not complete the picture. Some have argued that the Star Wars decision was driven, at least partially, by technological progress or the administration's confidence in technological solutions to strategic problems.[8] Bureaucratic politics might

---

4. Leslie H. Gelb in the *New York Times*, 25 March 1983, pp. 1, 8; and David Hoffman and Lou Cannon in the *Washington Post*, 26 March 1983, pp. 1, 7.

5. Many reports put an emphasis on the influence of Teller, including Robert Scheer, *With Enough Shovels* (New York, 1983); and William J. Broad, "The Birth of Star Wars," in *Claiming the Heavens* (New York, 1988), pp. 3–25.

6. R. Jeffrey Smith, "Reagan Plans New ABM Effort," *Science*, 8 April 1983, pp. 170–71; Scheer, *With Enough Shovels*, pp. 283–85 (taken from 10 July 1983 *Los Angeles Times* report). This view was then reiterated in subsequent analysis of Star Wars, for example, in George W. Ball, "The War for Star Wars," *New York Times Review of Books*, 11 April 1985, p. 39.

7. Hedrick Smith, emphasizing the secrecy and personal role of the president, sees the Star Wars decision as a precursor of the dysfunctional policy process of Iran-Contra; see his *The Power Game* (New York, 1988), p. 597.

8. Gerald Steinberg, "The Limits of Faith," in *Lost in Space: The Domestic Politics of the Strategic Defense Initiative*, ed. Gerald Steinberg (Lexington, Mass., 1988), pp. 145–58; Gregg Herken, "The Earthly Origins of Star Wars," *Bulletin of the Atomic Scientists*, October 1987, pp. 20–28.

have brought about the decision; the military service or services that stood to gain from such a program might have initiated the idea. Finally, the lobbying of industrial interests in aerospace and electronics may have inclined the Pentagon and the administration toward the program.

None of these several explanations, even in combination, adequately accounts for the nature and the timing of the decision. Although some of these factors are important as underlying causes, they also are underdetermining or insufficient. Only when domestic political competition is considered can a sufficient explanation be constructed. Attacks upon the administration's defense policies by the nuclear freeze movement and its allies, and threats to Reagan's popularity and political agenda as a whole, were the immediate and necessary provocations that gave rise to the decision. Domestic political competition appears to have determined both timing and character of the decision. The president and his White House advisors chose to implement, at the politically opportune moment, an option that had long been at their disposal.[9]

Strategic Defense Programs before Reagan

United States military space programs, including ballistic missile defense, have a long and expensive history that spans the cold war era. The military space race began in earnest during 1957, the year of Sputnik. From that time forward, the United States committed significant military resources to numerous space projects, including reconnaissance and warning satellites. The two general military objectives that have absorbed most of the resources are antisatellite (ASAT) weapons and ballistic missile defense (BMD). In the two decades preceding Reagan's announcement, the United States spent an average of $1.3 billion per year on missile defense programs.[10]

---

9. Daniel Wirls, "Defense as Domestic Politics: National Security Policy and Political Power in the 1980s" (Ph.D. diss., Cornell University, 1988). Two other accounts emphasize domestic pressures: Arthur Stein, "Strategy as Politics, Politics as Strategy," in *The Logic of Nuclear Terror*, ed. Roman Kolkowicz (Boston, 1987), pp. 186–210; and Janne Nolan, *Guardians of the Arsenal* (New York, 1989). While often emphasizing one or more influences most accounts do not attempt to weigh their relative importance.

10. BMD funding figures obtained from Congressional Budget Office, *Strategic Defenses*, June 1989, pp. 13–14. For a thorough account of the origins and history of the space race between the United States and the USSR see Walter MacDougall, *The Heavens and the Earth* (New York, 1985). A history of ABM/BMD strategy, technology,

Ballistic missile defense research began under Eisenhower with two programs. The first was an attempt to adapt the Nike interceptor missile, which was designed to destroy bombers, to the task of intercepting ICBM warheads. The second was Project Defender. Initiated in 1958, this top-secret program absorbed millions of dollars and employed thousands of scientists to explore new technologies for missile defense. One proposal, Bambi, or Ballistic Missile Boost Intercepts, involved space-based battle stations and rocket-propelled weapons designed to destroy missiles through direct impact. Elements of Bambi inspired some contemporary plans for comprehensive missile defense and closely parallel weapons that have actually been tested as part of the Strategic Defense Initiative.[11]

As early as 1959, scientists also began to investigate the missile defense potential of lasers, which were themselves just being invented. Alongside such basic research, work continued on ground-launched, nuclear-tipped missile interceptors, such as the army's Nike-Zeus. All along, work on ABM and ASAT development had been justified by reference to parallel Soviet developments and as a hedge against Soviet breakout in ASAT or missile defense.[12] Nike-Zeus, however, evolved into the Johnson administration's full-scale Sentinel program in 1967. Foreshadowing SDI, Sentinel was given many justifications, including comprehensive population defense, defense against attack by a small third power (i.e., China), or an inadvertent attack.[13] Nixon changed Sentinel into Safeguard and gave it the explicit goal of ICBM silo defense. At about the time that BMD development reached the production stage, the SALT negotiations curtailed the program with the ABM treaty of 1972. Not only is the ABM treaty the only arms control agreement to have prevented significant development and deployment of any weapons system,

---

and politics from 1955 to 1972 is given in Ernest Yarnella's *The Missile Defense Controversy* (Lexington, Ken., 1977). Paul Stares focuses on ASAT development in his *Space Weapons and U.S. Strategy* (London, 1985).

11. William J. Broad, "Star Wars Traced to Eisenhower Era," *New York Times*, 28 October 1986, pp. C1, 3. According to Broad, a 1982 High Frontier study of BMD "drew heavily on the Bambi proposal"; as we shall see, the High Frontier organization and its study became an inspiration for Reagan's 1983 decision. In 1984 the army and Lockheed tested the Homing Overlay Experiment, or HOE, which resembled Bambi in its use of a spreading metal web designed to increase the chances of impact.

12. Stares, *Space Weapons and U.S. Strategy*, pp. 174–75.

13. For a complete treatment of the McNamara-Johnson decision and the justifications for Sentinel see Morton Halperin, *Bureaucratic Politics and Foreign Policy* (Washington, D.C., 1974).

but it also precluded an arms race in a completely new strategic arena.

All this began to change in the late 1970s as the new cold war mobilization got under way. The nuclear "window of vulnerability" implied a need for some defense of ICBMs. Soviet civil defense programs and their renewed testing programs, especially of ASATs, indicated to analysts such as the members of Team B that the Russians were serious about defending themselves. Renewed Pentagon interest in military space programs began under Carter, during whose administration new programs began for both particle beam and laser beam research—both key elements in Reagan's SDI.[14]

The Reagan administration did not come in with public plans for a revitalized BMD effort, and the offensive strategic buildup at first obscured what governmental activity there was. But just over two years into his first term, Reagan suddenly announced that he was embarking on a new course in military policy: developing a strategic defense. While the announcement may have been sudden, conservative elites had for several years been reconsidering and promoting a revitalization of strategic defense.

The Rebirth of Elite Support for Strategic Defense

Even before the 1980 election, conservative politicians, activists, and scientists allied with Reagan were mobilizing support for strategic defense, particularly in the form of ballistic missile defense for U.S. ICBMs. The main justification for this renewed interest in BMD was the same argument conservatives offered for a new arms buildup in general: that the strategic balance was shifting in favor of the Soviets and leaving U.S. ICBMs vulnerable.[15] Particularly salient was the MX program and the possibility that BMD was the easiest and cheapest way to protect the new missile, especially when compared with Carter's multiple protective shelters plan. The need for a technological fix for the problem of land-based missile vulnerability was not, however, the only argument proffered for BMD. Gradu-

---

14. Stares, *Space Weapons and U.S. Strategy*, pp. 214–15.
15. Fred Kaplan, "Return of the ABM," *Atlantic Monthly*, September 1981, pp. 8–18; John Quirt, "Washington's New Push for Anti-Missiles," *Fortune*, 19 October 1981, pp. 142–48; Clarence A. Robinson, Jr., "Emphasis Grows on Nuclear Defense," *Aviation Week and Space Technology*, 8 March 1982, p. 27; Robert C. Gray, "Arms Control Implications of Ballistic Missile Defense," in *Antiballistic Missile Defense in the 1980s*, ed. Ian Bellamy and Coit D. Blacker (London, 1983), pp. 29–45.

ally, conservative analysts and politicians gave voice to positive endorsements of missile defense.[16] Strategic defense was not a necessary evil, but a potential escape from the inherent danger and immorality of Mutual Assured Destruction.

A small but important constituency for strategic defense formed on Capitol Hill. In 1979, Republican senator Malcolm Wallop of Wyoming published an article arguing that new technological opportunities (directed-energy weapons) pointed the way for a transformation of U.S. strategy from "assured destruction" to "assured protection" of American lives and homes. Wallop presented the article in 1980 to candidate Reagan, who read it and made written comments. Wallop also took his case to the Senate with an attempt to boost laser research funding by $250 million in the spring of 1981. In the summer of 1980 another stalwart of the New Right, Representative Jack Kemp (R-N.Y.), also lent early support to the idea of strategic defense.[17] Behind Wallop worked his Intelligence committee staffer, Angelo Codevilla. Codevilla tried without success to set up a strategic defense lobbying group, including industrial representatives. Corporate representatives were not opposed to an enlarged strategic defense program, but they were unwilling to lobby for it when it appeared that the Pentagon was unenthusiastic, at best, about such an effort.[18]

---

16. For examples from the strategic analysts see Colin S. Gray, "A New Debate on Ballistic Missile Defense," *Survival* 23 (March/April 1981): 60–71 (Gray cites the ICBM problem but goes on to opine that it is time "to reevaluate the wisdom of the offense-dominance."); and Stephen Peter Rosen, "Nuclear Arms and Strategic Defense," *Washington Quarterly* 4 (Spring 1981): 82–99.

17. On the evolution and activities of this Capitol Hill group see Angelo Codevilla, *While Others Build* (New York, 1988), chap. 4. Codevilla argues that much of the renewed interest stemmed from strategic intelligence activities, which used technology applicable to strategic defense. Malcolm Wallop's article is "Opportunities and Imperatives of Ballistic Missile Defense," *Strategic Review* 7 (Fall 1979): 13–21; in it he lends early support to a broad interpretation of the ABM treaty: that the treaty allows development of new technologies. On Reagan reading the Wallop article see David Hoffman, "Reagan Seized Idea Shelved in 1980 Race," *Washington Post*, 3 March 1985, pp. 1, 18. Wallop ended up getting fifty million dollars for laser research; his amendment was cosponsored by senators Helms, East, Symms, Hatch, Laxalt, Goldwater, Warner, Tower, Armstrong, Dole, Hayakawa, and a lone Democrat, Inouye. Wallop's (and Codevilla's) efforts are reported in Patrick E. Tyler, "How Edward Teller Learned to Love the Nuclear-Pumped X-ray Laser," *Washington Post*, 3 April 1983, pp. D1, 4; and Codevilla, *While Others Build*, chap. 4. Jack Kemp's article is "U.S. Strategic Force Modernization," *Strategic Review* 7 (Summer 1980): 11–17. Other early supporters included then-Representative Ken Kramer of Colorado and former Senator Harrison Schmitt of New Mexico.

18. Codevilla, *While Others Build*, pp. 89–91. Codevilla provides an example of Dan

As early as 1976, Reagan had mused about the lack of U.S. defenses against nuclear weapons, and his chief aides were aware of his concerns. In 1979, Martin Anderson, at the time in charge of policy development for the Reagan campaign, wrote several policy papers for the campaign. One of them, "Memorandum No. 3," discussed defense policy.[19] Anderson argued that Reagan could choose among three options on strategic military policy: sustain Carter's programs, match the Soviet buildup, or pursue strategic defense. Anderson's memo argued for the defensive alternative because it confronted simultaneously the Soviet threat and the political problem posed by the public's fear of war and an arms race. In addition, one of Reagan's military advisors in the 1976 and 1980 campaigns, Daniel O. Graham, a retired army lieutenant general and former head of the Defense Intelligence Agency, briefed the candidate in February 1980 on strategic defense as an "alternative to the all-offense MAD doctrine."[20] Senators Laxalt and Wallop and other congressional supporters spoke directly with the candidate on the subject. But others close to the candidate, including Michael Deaver, thought the idea politically dangerous as a campaign issue and kept it off the agenda.[21] Although he did not run on the issue, Reagan said during the campaign that a defense against nuclear attack was desirable if not necessary.[22] Moreover, the 1980 GOP platform, unlike the one in 1976, made an explicit endorsement of a revitalized strategic defense program.[23]

Tellep, then president of Lockheed Missiles and Space Company, who in December 1982 told Codevilla that his company would never "get ahead of the customer," meaning the Pentagon. Codevilla confirmed that support for strategic defense was growing "despite a lack of cooperation from industry . . . no lobbying" (interview with author).

19. Martin Anderson, *Revolution* (New York, 1988), pp. 85–86; see also Hoffman, "Reagan Seized Idea," p. 18.

20. Daniel O. Graham, *To Provide for the Common Defense* (copyright by Daniel O. Graham, 1986), p. 17.

21. Martin Anderson writes that the political advisors liked the "substance of the idea" but thought it vulnerable to Democratic attacks as a campaign issue. "In 1980 it was clearly an idea whose time had not yet come politically." Anderson, *Revolution*, pp. 86–87, and interview with author. General Robert Richardson, deputy director of High Frontier, interview with author; Hoffman, "Reagan Seized Idea," p. 18.

22. Scheer, *With Enough Shovels*, pp. 233–34. In this interview during the primary season of 1980, Reagan said it was ironic that we had no defense against nuclear weapons and that we ought to turn our expertise "loose on what we need in the line of a defense against their weaponry and to defend population."

23. As one of its strategic priorities the platform said the Republicans would "proceed with . . . vigorous research and development of an effective anti-ballistic missile

Strategic Defense in the Reagan Military Program before 1983

Although attention after Reagan's inauguration focused on the offensive strategic buildup, the Department of Defense was reshaping research programs on the military use of space. These programs received increased funding along with everything else in the Pentagon budget. Despite some inclinations toward aspects of strategic defense, no major initiative was planned. In the summer of 1981 the president ordered NSC and Pentagon reviews of military space policy. Most of the administration's initial plans for military space programs centered on communications, warning, navigation, and especially ASAT weapons programs, while BMD was mentioned only as a possible solution to the problem of ICBM vulnerability.[24] New bureaucratic organizations were created to match the growing emphasis on space programs.[25] During his confirmation hearings to be presidential science advisor, George Keyworth, fresh from his position at Livermore Laboratory, where he worked with Edward Teller, testified that laser and directed-energy weapons "represent an enormous possibility for the future . . . they represent the only truly credible antiballistic missile alternative for the future." He cautioned, however, that such technologies were years from fruition.[26] Keyworth made the formation of a national task force to direct BMD research a high priority—an idea that foreshadowed the Strategic Defense Initiative Organization, or SDIO. Secretary of Defense

---

system, such as is already at hand in the Soviet Union, as well as more modern ABM technologies." *National Party Platforms of 1980*, comp. Donald Bruce Johnson (Urbana, Ill., 1982), p. 207. Richard Allen, who was advising the Reagan campaign and would become the president's first national security advisor, was an important figure in the drafting process at the convention.

24. This is evident in yearly reports such as the annual reports to Congress by the secretary of defense and the military posture statements issued by the Joint Chiefs. For more details see Philip M. Boffey, "Pressures Are Increasing for Arms Race in Space," *New York Times*, 18 October 1982, pp. 1 and B8; and Richard Halloran, "U.S. to Increase Military Funds for Space Uses," *New York Times*, 29 September 1982, pp. 1 and 24. The debate over ASAT weapons preceded SDI and remained an important controversy after Star Wars was initiated; see Larry Pressler, *Star Wars* (New York, 1986).

25. Stares, *Space Weapons and U.S. Strategy*, pp. 219–20. In addition to promoting the North American Air Defense commander in chief to four-star rank, the air force started a space command equal in stature to other air force commands. The DOD also started a space operations committee.

26. *Aviation Week and Space Technology*, 27 July 1981, p. 26.

Weinberger gave an early indication of growing support for comprehensive strategic defense in a 1981 interview. He stated that "we want both offensive and defensive strategic strength" and that defensively, "what we would like to have is something that makes a ballistic missile attack on the U.S. fully ineffective."[27]

Official documents drafted before the 1983 decision provide ambiguous indications about the administration's commitment to strategic defense. The defense secretary's FY 1983 annual report to Congress is quite cautious about BMD technology but notes that the Pentagon plans to pursue a research program to provide defense for ICBMs.[28] Strategic defense has its place in the defense guidance statement of 1982, which discusses the possible development of area or population missile defense and point defense for the MX in particular.[29] Even in the FY 1984 annual report, completed on 1 February 1983, BMD is listed last in the discussion of "strategic defense forces," including warning systems, interceptors, and ASAT. The 1984 report also provides a table that shows a sharp rise in planned development funds for BMD from $426 million in actual funding in 1982 to a proposed $519 million in 1983, $709 million in 1984, and $1,564 million in 1985.[30] The $1.56 billion projected for 1985 comes quite close to the actual figure requested for that year, after the organization of SDI in late 1983.

Despite some favorable statements and increased budgets, the Department of Defense was planning no initiative in strategic defense. Just hours before Reagan delivered the Star Wars speech, air force major general Donald Lamberson, assistant for directed-energy weapons, stated in congressional testimony regarding strategic defense programs, "Our goals . . . are rather modest."[31] In fact, up until the decision in March, Defense and State Department officials and technical studies by the defense bureaucracy cast significant doubt on the technical feasibility, affordability, and strategic ad-

---

27. Quirt, "Washington's New Push for Anti-Missiles," pp. 142 and 145.
28. *FY 1983 Annual Report*, 8 February 1982.
29. Richard Halloran, "Pentagon Draws Up First Strategy for Fighting a Long Nuclear War," *New York Times*, 30 May 1982, p. 12.
30. *FY 1984 Annual Report*, 1 February 1983, p. 228.
31. "The Development of Defense Directed Energy Program and Its Relevance to Strategic Defense," statement by the assistant for directed-energy weapons, U.S. Senate, Committee on Armed Services, Subcommittee on Strategic and Theater Nuclear Forces, 98th Cong, 1st sess., 23 March 1983, p. 5. Cited by G. Allen Greb, "Short-circuiting the System," in Steinberg, *Lost in Space*, p. 33.

visability of a defensive system. In 1987, Senator Bennett Johnston obtained copies of Pentagon reports done in 1982 on the High Frontier proposal, one of the inspirations for the Star Wars initiative.[32] One joint air force–army project under the auspices of the Air Force Systems Command's Space Division is dated 31 March 1982, just a year before the Star Wars speech. Drawing upon a large group of officials from the air force, army, and industry, the report concludes that the High Frontier "concept, as proposed, is not technically feasible" and that the proposal should "not be funded as proposed, nor modified and funded."[33] In correspondence about the High Frontier study, Weinberger wrote High Frontier's founder, Daniel Graham, that "we are unwilling to commit this nation to a course which calls for growing into a capability that does not currently exist."[34] State Department evaluations reached conclusions similar to the Pentagon's. In November 1982, Richard DeLauer, undersecretary of defense for research and engineering, told a congressional committee that an ABM program would be "a multiple of Apollo programs" and that Congress would be "staggered at the cost."[35] Even the White House Science Council, just two weeks before the Star Wars speech, produced a report that "did not foresee any scientific breakthroughs in the next ten years and did not call for a major stepped-up effort."[36] No department within the Pentagon was pushing for a dramatic initiative in strategic defense; indeed, most were arguing against any such effort.[37]

This barrage of skepticism from the bureaucracies responsible for strategic defense programs appears to belie claims that the presi-

32. The High Frontier study proposed using kinetic-energy weapons based on satellites. Elements of the High Frontier concept are very much part of the Phase I SDI deployment planned for the mid-to-late 1990s. Senator Johnston's efforts and the substance of the studies were first reported by William J. Broad, "Space Weapon Idea Now Being Weighed Was Assailed in 1982," *New York Times*, 4 May 1987, pp. 1, B20.
33. "Memorandum for Record" on "Evaluation of SRI International's Global Ballistic Missile Defense Unsolicited Proposal," dated 31 March 1982, and obtained from a congressional office.
34. Letter from Caspar Weinberger to Daniel Graham, dated 24 November 1982, and obtained from a congressional office.
35. Quoted in Michael R. Gordon, "Reagan's 'Star Wars' Proposals Prompt Debate over Future Nuclear Strategy," *National Journal*, 7 January 1984, p. 16.
36. Ibid., p. 15, and Broad, "Birth of Star Wars," p. 18.
37. Angelo Codevilla provides several instances of rather active resistance by various Pentagon officials to the efforts of the Capitol Hill laser lobby; see *While Others Build*, pp. 71–88.

dent's decision was based on technological progress made in the few years preceding the decision.[38] While everyone acknowledged the considerable technological progress since the first ABM debate, no research agency within the Pentagon saw a development or change in capabilities that would make strategic defense effective and affordable.

The White House Network: Keeping the Idea Alive

While the administration had no public plans for a major initiative in strategic defense, and while the Pentagon was reacting with skepticism to such proposals, the idea of strategic defense was kept alive, not only by Wallop and his allies on Capitol Hill, but also by a small network of White House officials and private elite activists. At the center of this network were four of the president's closest advisors: Richard Allen, national security advisor; Martin Anderson, assistant to the president for policy development; presidential counselor Edwin Meese; and George Keyworth, the science advisor.[39] This core group cooperated closely with Reagan's kitchen cabinet of private businessmen, including Karl Bendetsen, Joseph Coors, William Wilson, and Jacquelin Hume, who were all supportive of strategic defense. Finally, working one step outside these inner circles were the most public advocates of strategic defense, General Graham, and later his organization, High Frontier, and Edward Teller, father of the H-bomb and director of Livermore Laboratory.

Even before Teller had seen the president, or Graham had finished the influential High Frontier study, White House advisors were discussing strategic defense among themselves and with members of the kitchen cabinet. According to Martin Anderson, the first meeting took place on 14 September 1981 in Meese's office and included Anderson, Bendetsen, Graham, Teller, and Keyworth. The

---

38. George Keyworth stated, "What made him [Reagan] decide that 'now' was the time to begin this initiative was the state of technology and the rate of progress, in the last few years, underlying those technologies." Yet less than two years before the Star Wars announcement, Keyworth had testified, in his confirmation hearings, "I would caution that both the U.S. and the Soviets are a great distance away from being able to deploy a large high-energy laser system by any means capable of achieving an ABM objective." From Michael Charlton, *From Deterrence to Defense: The Inside Story of Strategic Policy* (Cambridge, Mass., 1987), pp. 102–3; and *Aviation Week and Space Technology*, 27 July 1981, p. 26, respectively.

39. The following account of this network's actions is drawn largely from Anderson, *Revolution*, pp. 92–96, and from interviews with Anderson and Meese.

eventual product of that meeting was a report by the outside advisors on the feasibility of strategic defense. They presented the report privately to the president on 8 January 1982 with only members of the network, now including the new national security advisor, William Clark, in attendance. Despite the president's satisfaction with the report, no decision to proceed was made, and the issue lay dormant in the White House for almost a year.

During 1982 the most visible and vocal proponent of strategic defense was Daniel Graham.[40] Graham and his colleagues were organizing a major study of the technical feasibility, military soundness, economic viability, and political acceptability of a strategic defense system. This ad hoc effort was supported financially by members of the kitchen cabinet and by the Heritage Foundation.[41] One result was the formation of High Frontier as an independent public interest group and the publication in March 1982 of that seminal work on space-based defense in the 1980s, *High Frontier: A New National Strategy*, which I have been calling the "High Frontier study."[42] Just before the study's publication, those of its sponsors who were also members of Reagan's kitchen cabinet, including Bendetsen, Coors, and Hume, presented its findings to the president.[43] The study recommended rapid (two to three years) deployment of point defense for ICBM silos; deployment of a space-based BMD system in five to six years, and development and deployment of an advanced-technology space defense system within ten to twelve years. Graham's study proposed using largely off-the-shelf technology to build kinetic-energy weapons to destroy missiles. Teller, meanwhile, was pushing more exotic X-ray lasers, driven by a nuclear explosion.

---

40. Graham had introduced his ideas in a 1981 article in which he praised the administration's plans for strategic and conventional modernization. In it, he warned, however, that "even massive infusions of money, poured into the old categories of military programs, could eventually leave the United States in a strategically worse position relative to the Soviet Union." The discredited offensive policy of MAD, needed, he urged, to be replaced by a "new strategic framework," embodied in a space-based BMD. Graham added cost effectiveness to the list of BMD benefits: defense would cost less in the long run than an interminable offensive nuclear arms race. Daniel O. Graham, "Toward a New U.S. Strategy," *Strategic Review* 9 (Spring 1981): 9–16.

41. Kitchen cabinet financiers included Bendetsen, Coors, and Justin Dart. High Frontier tried to organize under the American Security Council (the group so important in the SALT II campaign) but did not for financial reasons. General Richardson, interview with author.

42. Daniel O. Graham, *The Non-Nuclear Defense of Cities* (Cambridge, Mass., 1983), which is another edition of the original High Frontier study.

43. Daniel O. Graham, *To Provide*, p. 17.

This difference in technological commitments eventually led Teller to distance himself from High Frontier, after having served on its advisory board.

While strategic defense lay dormant in the White House during much of 1982, High Frontier spearheaded an effort to mobilize public opinion for strategic defense from June to the next spring. With the help of the Conservative Caucus and its local branches, High Frontier representatives gave presentations in forty states "to every . . . multiplier of public opinion you could find."[44] The effort resulted in several favorable editorials across the country. More important, according to High Frontier's General Robert Richardson, this campaign generated about one hundred letters a day to the White House and the Pentagon in support of strategic defense during the weeks before Reagan's March address. In effect, High Frontier floated a private-sector trial balloon to demonstrate to the administration the potential popularity of such an initiative. Richardson claims that by the time the president was set to deliver the Star Wars message, the White House, which had been monitoring the public campaign, "knew within one percentile point what the public reaction would be."[45] Although it is not clear whether the White House was tracking its own polls, some pollsters were asking the public about strategic defense before 1983 and finding opinion to be favorable. For example, Sindlinger and Associates found in August 1982 that when asked "if the U.S. had the capability of changing this situation [that we have no defense against ballistic missiles] by deploying an antiballistic missile defense, would you favor it being done?" 86 percent responded affirmatively and only 10 percent negatively.[46]

While Edward Teller and Graham's High Frontier were important public advocates, it was White House advisors who guided and directed the movement toward strategic defense. Meese, Anderson, Allen (and later Clark) initiated the earliest meetings on the issue and regulated access to the president on the topic. Teller and Graham did not introduce the concept of strategic defense into the political vocabulary of Ronald Reagan and his advisors. Upon assuming the presidency, Reagan was already "philosophically" convinced

---

44. General Robert Richardson, interview with author.
45. Ibid.
46. Poll results reproduced by Thomas W. Graham and Bernard M. Kramer, "The Polls," *Public Opinion Quarterly* 50 (Winter 1986): 129.

that strategic defense was desirable, but he took no decisive action on the issue for two years.[47] Considering the firm skepticism emanating from official bureaucratic channels, including direct rejection of the High Frontier approach, which clearly was more feasible than Teller's lasers, the White House seemed to be looking for scientific credibility to justify its strategic inclinations. Nevertheless, despite widespread agreement by early 1982 among White House staff and the president on the desirability of a strategic defense system, no such decision was made for nearly a year.

The Context and Timing of the Star Wars Decision

Given the lack of an industrial lobby or bureaucratic pressure, no breakthrough in technology, and lingering indecision in the White House, other factors would seem necessary to account for the sudden Star Wars initiative. Martin Anderson argues that, besides technical questions, political concerns determined when the president would go ahead with the initiative.[48] Indeed, the initiative was not forthcoming until certain political circumstances coalesced, though not the ones Anderson describes. Pivotal were the efforts of the nuclear freeze movement and the opposition it spawned in Congress to elements of the buildup, particularly the MX. The peace movement seemed at the end of 1982 to be riding a tidal wave of public support and political success. It had, after all, sponsored the largest single political rally in U.S. history and almost gotten the House to pass the freeze resolution in August. The arms race and the freeze had figured prominently in the 1982 elections. By December, the movement had forced Congress to paralyze the MX program, the centerpiece of the strategic modernization program. And support for an arms control agreement based on a nuclear freeze continued to gain momentum. At the start of 1983 more than 70 percent of the public, with solid majorities in every demographic category, favored a bilat-

---

47. Both Meese and Anderson took this position in interviews with the author. At most, Teller and Graham can be credited with making strategic defense appear to be technically feasible to White House staff or with providing them with enough technical ammunition for such a proposal.

48. "Reagan knew it was time to go, politically speaking. It was not the perfect time, but it was likely to be the best time he would ever have. He was settled in his job, his respect at home and abroad was growing, and his power as a sitting president about to run for reelection was nearing its peak" (*Revolution*, p. 98). As we shall see, Anderson correctly points to political timing but is rather amiss in his analysis of the state of the president's power.

eral and verifiable freeze on testing, production, and deployment of nuclear weapons; even among those who thought such an agreement unverifiable, the margin was two to one in favor of a freeze.[49] The offensive military buildup was losing its public appeal, especially given the recession and the deficit. When it came to reducing the deficit, which had become a critical issue for both parties, by January 1983, 57 percent of the public approved of cuts in the defense budget to offset the deficit, while 41 percent opted for reducing social programs, 18 percent for higher taxes, and 12 percent for reducing entitlement programs.[50] Reagan's defense programs seemed destined for significant cutbacks. Aggravating the administration's problems was the severe recession of 1982, which also made life difficult for Republicans during the midterm elections. Together these circumstances pushed Reagan's job approval rating to the nadir of his presidency: from 49 percent in January 1982 to 35 percent by late January 1983.[51] The administration was facing a political crisis. Contrary to the statements of some administration officials, the stalemate that inspired Star Wars was less strategic than it was domestic, less an impasse with the Soviets and the offensive arms race than with Congress and the peace movement.

Advocates of strategic defense and others realized that the threat the peace movement posed both required a response and provided a political rationale for strategic defense. General Graham put this argument foremost in his 1982 letters on the High Frontier concept to Secretary Weinberger. In one dated 31 March 1982, Graham begins: "In light of the current country-wide press for 'nuclear freeze,' I wish to bring to your attention the value of the High Frontier concept as an effective counter." In a later letter he states that the "concepts we endorse constitute the best currently available riposte to the new surge toward 'nuclear freeze,' 'no-first-use' and other related proposals."[52] In a mid-1982 issue of *Conservative Digest*, Gregory Fossedal of the Heritage Foundation said that the administration had a plan for a satellite antimissile system, which he said would

49. *Gallup Reports* 208 (January 1983): 13.
50. Ibid. 209 (February 1983): 16–17.
51. In May of 1981, approval of the way Reagan was handling his job was 68 percent positive and 21 percent negative. From then, positive ratings fell steadily through the 40 percent range as the negative ratings rose during 1982. In January of 1983, approval reached an all-time low at 35 percent, with disapproval running 56 percent. Ibid. 213 (June 1983): p. 28.
52. Letter dated 14 April 1982. Both letters from Graham to Weinberger are from material released to Senator Bennett Johnston.

The Strategic Defense Initiative

make arms control negotiations irrelevant and could be "the secret weapon . . . to undercut the freeze crusade."[53] In her testimony during congressional hearings on the freeze proposal in early 1983, Phyllis Schlafly stated, "American technology has developed an 'anti-freeze' defensive system that can . . . prevent war and save lives. Nuclear freeze, on the other hand, can't do either one."[54] Even the skeptical air force study of the High Frontier proposal cited earlier had one positive conclusion: "The study does make a very useful contribution. It raises the question whether our national strategy needs revision. 'Business as usual' will not satisfy the body politic—witness the sharp decline in public support for defense within the last year and the increased concern over nuclear weapons."[55]

Indeed one of the most important manifestations of the decline in support for defense, the congressional restriction of funds for the MX in late 1982, precipitated meetings and decisions in the White House that culminated in the Star Wars speech. In December 1982, William Clark arranged a meeting between Reagan and the Joint Chiefs of Staff, during which the president, according to one account, made a vague, almost rhetorical, statement about moving toward defense.[56] After the meeting, one of the Joint Chiefs asked Clark whether Reagan's remark meant they should study the issue. Clark replied affirmatively and then asked his deputy Robert McFarlane to investigate the defense option. Several weeks later, on 11 February 1983, Reagan met with staff members and the Joint Chiefs.[57] While the MX dilemma was the central topic, strategic defense, which was one option for salvaging the MX and perhaps the strategic buildup as a whole, became a focus of the meeting. According to some reports, Clark and McFarlane arranged to have the topic raised by the Joint Chiefs.[58] Admiral James D. Watkins, chief of na-

---

53. Reported in *Nuclear Times*, October 1985, p. 5.
54. U.S. House, Committee on Foreign Affairs, *Calling for a Mutual and Verifiable Freeze on and Reductions in Nuclear Weapons*, 98th Cong., 1st sess., 17 February, 2 and 8 March 1983, p. 402.
55. "Assessment of High Frontier Study: A Summary," part of the air force evaluation of the High Frontier proposal, obtained by Senator Bennett Johnston.
56. Without citation, Anderson quotes Reagan as asking, "What if we began to move away from our total reliance on offense . . . toward a relatively greater reliance on defense?" *Revolution*, p. 97.
57. On this meeting see Hedrick Smith, *Power Game*, pp. 607–9; Frank Greve, "'Star Wars': How Reagan's Plan Caught Many Administration Insiders by Surprise," *San Jose Mercury News*, 17 November 1985, p. 21; Hoffman, "Reagan Seized Idea," p. 18; Herken, "Earthly Origins of Star Wars," pp. 24–25; Broad, "Birth of Star Wars," p. 20.
58. Hedrick Smith, based on interviews with most of the principals, including

val operations, presented the case for a movement toward defense, based on his "Freedom from Fear" paper.[59] Nevertheless, no explicit decision was reached at the meeting, and the endorsement of strategic defense by the Joint Chiefs was vague and unspecified.[60]

After the meeting, Clark delegated to McFarlane the task of putting something together on strategic defense. It was the upcoming presidential address in defense of the embattled military program, however, which drove the decision-making process. Reports said Reagan's political advisors wanted something different and upbeat for the speech on defense.[61] After receiving some negative press on his 8 March speech, during which he characterized the Soviet Union as the "focus of evil in the modern world," the president himself reportedly expressed an unspecified desire for something new and positive to say about defense.[62] Reportedly, it was only then that Clark and McFarlane decided to include strategic defense in the upcoming address.[63] The great secrecy surrounding McFarlane's work has been a favorite theme of Star Wars historians. These accounts dwell on the exclusion of cabinet members, the Joint Chiefs, and other high-ranking Defense and State Department officials from the

---

McFarlane, describes "backchannel" recruitment of Watkins by McFarlane (*Power Game*, p. 607). Greve says that the JCS was stymied by the stalemate with Congress and had nothing new to offer the president. General John W. Vessey called the Joint Chiefs together before the meeting because they, according to a participant, "wanted to bring the President something new, different, and exiting"; quoted in "'Star Wars,'" p. 21.

59. Representing the service least likely to gain from strategic defense, Admiral Watkins was reportedly more motivated by his devout Catholicism and the impact of the bishops' critique of deterrence.

60. There is a consensus among those interviewed by the author and most other researchers that the JCS presentation was general and devoid of specific recommendations and that no decision was made to make defense an immediate priority. As General Vessey later said of the meeting's conclusion, "There was no program definition. . . . It was the idea that defense might enter the equation more than in the past"; quoted in Hedrick Smith, *Power Game*, p. 609. See also Greve, "'Star Wars,'" p. 21.

61. Lawrence Barrett, "How Reagan Became a Believer," *Time*, 11 March 1985, p. 16; Hedrick Smith, *Power Game*, p. 610. As indicated earlier, a number of insiders thought the speech was a way of "diverting attention from the nuclear freeze movement"; see Gelb, "Aides Urged Reagan," p. 1.

62. Herken writes that some time after the speech, Reagan reportedly told Clark and McFarlane that he "was reluctant to repeat the same litany of bad news in the threat speech . . . he said it should be supplemented with a more positive and compensating vision"; see "Earthy Origins of Star Wars," p. 25. The mixed reaction to the speech indicated the dwindling effectiveness of the theme.

63. Greve reports that it was after this that the NSC staff working on strategic defense got orders to produce an ending for the upcoming speech; see "'Star Wars,'" p. 21.

formulation of the speech insert and the decision to include it.[64] While the pending announcement was a well-kept secret, it is not true that none but a few NSC staff knew it would be part of the defense speech. Missing from these accounts is the fact that Reagan's closest nonmilitary advisors knew about the plans for the speech.[65] The extent to which the process was coordinated between White House national security aides and other political advisors remains somewhat cloudy, however.

Regardless of the level of coordination, similar motivations guided both national security and political staff. Clark and McFarlane were under political rather than strategic pressure to come up with something different, and strategic defense appeared to be the answer. As for the president, McFarlane said, "Reagan's view of the political payoff was sufficient rationale as far as he was concerned . . . providing the American people with an appealing answer to their fears."[66] Although McFarlane may have written the speech insert secretly, he did so at the direction of Clark. In fact Clark reportedly was the true enthusiast. He was a political confidant of Reagan's, not a military expert, and he was as much an insider as any of the troika of Meese, Deaver, and James Baker. Thus Reagan and his political advisors, who were in need of a new but unspecified initiative and who were fully aware of the strategic defense option, had that need fulfilled by White House military advisors, who were seeking remedies for a domestically produced crisis in defense policy. This led to a dramatic presentation of the idea with little preparation, despite the years of prior work on the subject.

Weighing Alternative Explanations

The best explanation for the sudden announcement of the strategic defense initiative is, then, that the administration chose to implement the idea in response to domestic political circumstances. Star

---

64. For accounts of who was excluded and when they were informed or surprised see Greve, "'Star Wars,'" p. 1; see also Hedrick Smith, *Power Game*, pp. 612–15. Among those excluded were Weinberger, Schultz, the Joint Chiefs, and the undersecretaries of state and defense including Richard Perle and Richard DeLauer. Some State Department and Pentagon officials did object to the initiative when notified and tried either to delay or modify the speech.

65. Edwin Meese, interview with author. Meese ascribed part of the press's emphasis on the secrecy to the fact that the announcement was not leaked, which was unusual.

66. Hedrick Smith, *Power Game*, p. 609 (from interview of McFarlane by Smith).

Wars emerged from the political process in the White House. Other explanations do not bear up under scrutiny. Very little evidence exists to support an explanation based on economic interests and pressures, given how efforts to mobilize an industrial lobby on behalf of strategic defense had failed. Less evidence can be marshaled for a bureaucratic politics explanation, when no military service had made a special case for strategic defense, and defense bureaucracies had issued skeptical reports and even resisted congressional efforts to add to programs. There had been no technological breakthroughs between 1980 and 1983, and none were foreseen. Technical progress was ongoing, but that does not explain why the White House circumvented pessimistic bureaucratic channels to hear optimistic reports by outsiders, nor does it explain the timing or suddenness of the announcement. Technical progress was a necessary but not a sufficient reason to launch Star Wars.

Rational actor explanations do not stand up either. Certainly Reagan's advisors knew well before 1983 that the cold war could not be won through an offensive nuclear arms race. The strategic stalemate with the Soviets was nothing new. No significant development in Soviet offensive power suggested the need for a comprehensive strategic defense.[67] The MX was the intended solution to the problem of vulnerability. Furthermore, at the same time the decision to announce Star Wars was being made, the President's commission on Strategic Forces, better known as the Scowcroft commission, was fashioning a long-term solution to the vulnerability problem that in no way involved BMD, let alone an astrodome defense. Impaneled to do a comprehensive review of all aspects of strategic policy, the commission did not endorse a shift toward strategic defense and drew skeptical conclusions about its potential. While the commission did not completely discount the ICBM window of vulnerability, they did say that it was not an imminent threat and "is not a sufficiently dominant part of the overall problem of ICBM modernization to warrant other immediate steps being taken such as . . . ABM defense of those silos." The report goes on to state that despite progress, "applications of current technology offer no real promise of being able to defend the United States against massive nuclear at-

---

67. Besides Reagan's reference to a long-standing "awesome" threat of Soviet ICBMs, Robert McFarlane has stated that one of the factors that swayed him toward strategic defense was "the unique qualities of the new Soviet systems coming on line," especially mobile ICBMs (Herken, "Earthly Origins of Star Wars," p. 24). But, again, mobile missiles in no way dictate a defensive response.

tack in this century" and that even point defense of missile silos would remain problematic. The commission, moreover, discounted the potential threat posed by Soviet strategic defense programs. Although it cited the Soviet ABM effort and called for a rigorous research and development program, the report did not mention defense in its section "Preventing Soviet Exploitation of Their Military Programs." Before March 1983, official Pentagon "threat" documents emphasized the Soviet air defense program, not their BMD programs. In fact, air defense comprised the bulk of the more than two hundred billion–dollar Soviet "Red Shield" Reagan was fond of invoking in defense of SDI. Only after 1983 did references to Soviet antimissile activities proliferate. The Soviet strategic defense program was largely an after-the-fact justification.[68]

The lack of rational planning is also evident from the fact that there was no coordination between the two efforts even though the Scowcroft commission was drafting its recommendations as the Star Wars speech was being written.[69] This distinct lack of relationship and coordination points once again to the paramount importance of domestic politics. The principal recommendation of the commission—to place the MX in fixed silos and proceed with development of a small, single-warhead missile—did almost nothing to end the debate over ICBM modernization but served its immediate purpose of saving the MX in Congress. The primary target of the Scowcroft commission was the House of Representatives. The objective of the Star Wars speech was to influence public opinion. Star Wars was intended to provide the administration with a positive initiative in the realm of strategic and defense policy at a time when it was losing the public relations battle in that area. Scowcroft would save the MX; Star Wars would undermine some of the public concerns and opposition engendered by the nuclear freeze campaign. The domestic objectives, rather than any strategic designs, make sense of these two major but utterly uncoordinated initiatives. Even if the administration was partly motivated by a sincere desire to circumvent MAD or to reestablish military superiority, neither motivation can explain the timing and character of the decision and the announcement.

---

68. Indeed, according to a major 1988 congressional study of SDI, based on interviews with SDI researchers and intelligence specialists, by the U.S. definition, "there is no Soviet SDI program as such"; James T. Bruce, Bruce W. MacDonald, and Ronald L. Tammen, *Star Wars at the Crossroads*, staff Report to Senators Bennett Johnston, Dale Bumpers, and William Proxmire, 12 June 1988, pp. 4 and 102–3.

69. Most sources agree that there was no connection between the efforts.

Domestic political concerns appear to have been decisive; even principals such as McFarlane admit to some connection. Without the pressures of the peace movement and Congress, it is probable that a strategic defense program would not have been formulated and announced, and certainly not at that juncture. Both the abruptness of the announcement and the boldness of the vision seem inextricably linked to efforts to meet political needs. Domestic politics qualifies as the immediate and necessary cause by explaining critical aspects that other largely underlying causes leave insufficiently determined.[70] The only new threat to Reagan's military programs was from domestic forces. Star Wars was less a technological end run around the Soviet offensive threat than a political end run around the peace movement and its political allies, who threatened the administration's power.[71] Thus the very success of the peace movement was responsible, in large measure, for Star Wars.

From Presidential Initiative to Pentagon Program

After the initial shock effects from the speech, the political impact of Star Wars grew gradually as it was transformed from a presidential vision into a Pentagon program: the Strategic Defense Initiative. The 23 March speech precipitated an immediate flurry of press reports and floor statements by members of both houses of Congress, which evinced further polarization of the strategic debate. Much like the freeze, Star Wars drew ridicule from its opponents and righteous praise from its advocates. But after the initial coverage and reaction, the issue dropped off the charts until late in the year, when Reagan's directive emerged from the Pentagon in the form of a five-year, twenty-six billion–dollar research program, with a deployment decision to be reached around 1990. Ironically, the strategic and technical planning process that many thought should have preceded the president's announcement was organized and used, instead, to sell the already formalized inititative. Two ad hoc commissions

---

70. Omnipresent underlying causes often cannot explain why something happened at all, let alone why it happened when it did or the way it happened. Immediate causes are often crucial to understanding exactly the how and the why. For a brilliant application of the distinction between underlying and immediate causation in international crises see Richard Ned Lebow, *Between Peace and War* (Baltimore, 1981).

71. "Technological end run" was a phrase used by High Frontier and other strategic defense advocates.

served this purpose. One panel, headed by James Fletcher and called the Defensive Technologies Study Team, studied the technological feasibility of strategic defense, while the Future Security Strategy Study panel, headed by Fred Hoffman, studied the strategic implications of defense. Although the Fletcher and Hoffman studies did not provide a ringing endorsement of the strategic wisdom or the technological feasibility of a comprehensive defense, the Pentagon used the joint findings to bolster its case for the SDI.[72]

Reagan quickly converted what had been a short-term public relations effort and a presidential vision into the top priority in his defense program and the most expensive project in the defense budget by FY 1987.[73] Beginning with the FY 1985 budget, SDI research funding began a rapid ascent. Despite cuts imposed by Congress every year, the level of spending increased dramatically, tripling in just three years. The budget figures in Table 11 do not include the more than three hundred million dollars a year in SDI research done by the Department of Energy, which adds approximately another 10 percent to total SDI funding.

Despite scientific, political, and budgetary controversies that surrounded the program, SDI developed, grew, and prospered. Four

Table 11. SDI funding, 1984–89 (in billions of 1990 dollars of budget authority)

	1984	1985	1986	1987	1988	1989
DOD SDI funds ($)	1.2	1.6	3.1	3.7	3.9	3.8
Real growth (%)		33.3	93.8	19.4	5.4	−2.6
SDI as percentage of total military R&D	3.7	4.3	8.1	9.4	10.0	9.8

Source: Congressional Budget Office, based on data from the Department of Defense.

72. The Fletcher study, titled "The Strategic Defense Initiative Defensive Technologies Study," is reprinted in U.S. Senate, Committee on Foreign Relations, *Strategic Defense and Anti-Satellite Weapons*, 98th Cong., 2d sess., 25 April 1984, pp. 141–75. The Hoffman study, titled "Ballistic Missile Defenses and U.S. National Security, Summary Report," is reprinted in the same volume, pp. 125–40. In general, SDI advocates cited the technical optimism of the study, while critics argued that the reports backed away from the president's plan because of its probable infeasibility. For example, contrast Clarence A. Robinson, Jr., "U.S. Strategic Defense Options: Study Urges Exploiting of Technologies," *Aviation Week and Space Technology*, 24 October 1984, pp. 50–51, with R. Jeffrey Smith, "Star Wars Panel Highlights Uncertainty," *Science*, 6 April 1984, p. 33.

73. Bill Keller, "No. 1 Weapon in the 1987 Budget Is Missile Shield," *New York Times* 5 February 1986, pp. 1 and 22. In his *FY 1988 Annual Report*, p. 4, Weinberger lists SDI as the "most important" defense program.

policies helped to accomplish this institutionalization. First, the administration nourished a proliferation of justifications, strategic and otherwise. Second, it acted quickly to maximize the budgets for SDI and the rapid distribution of those funds to military bureaucracies and private interests. Third, Reagan removed Star Wars from the bargaining table of arms control negotiations. Fourth, the administration pushed toward an early deployment decision. The reason for these policies, according to Attorney General Meese, was "so that it [a strategic defense system] will be in place and not tampered with by future administrations."[74]

Strategic Justifications

The overt justifications for SDI multiplied as rapidly as the program expanded. In the beginning, there was Reagan's vision of a world freed from the threat of nuclear war. Star Wars would substitute defense for deterrence. President Reagan, Caspar Weinberger, and Lieutenant General James A. Abrahamson, the director of SDIO, repeatedly stated that a fully effective comprehensive defense remained the ultimate goal of the program. In his FY 1988 annual report, Weinberger stated that the object of SDI was to "secure a thoroughly reliable defense against Soviet nuclear missiles to protect all our people."[75] When scientific criticism cast doubt on this ultimate goal, some in the bureaucracy and the administration immediately began to downplay the president's vision and talk instead about the enhancement of deterrence. This became the preferred argument for many administration and Pentagon officials, especially after the Fletcher and Hoffman studies made the potential of defense to strengthen deterrence a principal focus. General Abrahamson, in Senate testimony in early 1984, stated, "First and foremost, an effective defense against ballistic missiles would improve stability and reduce the likelihood of war by eliminating the military utility of a preemptive nuclear strike."[76] A defensive system of even marginal

---

74. Quoted in Council for a Livable World, *"Reagan Is Trying to Perpetuate His Military Policies,"* pamphlet (N.p., May 1987).
75. *FY 1988 Annual Report*, p. 5. Less than a week after the March 1983 speech, Weinberger confirmed that the president sought a defense that is "thoroughly reliable and total"; quoted by David Hoffman, "Futuristic Soviet Defense Said Welcome Possibility," *Washington Post*, 28 March 1983, p. 5. See also "Weinberger Denies Antimissile Shift," *New York Times*, 6 September 1986, p. 9.
76. U.S. Senate, Committee on Foreign Relations, *Strategic Defense and Anti-Satellite*

efficiency would prove valuable to the extent that it would increase the Soviet Union's uncertainty about the probable success of an attack, especially if SDI were used to protect only U.S. strategic forces and command centers, a far less formidable task that does not require a flawless system. In this way, the president and Weinberger would articulate the public vision of a world free from the threat of nuclear war while the generals and bureaucrats would offer more realistic plans during congressional testimony. Another argument reminiscent of the 1969 ABM debate was that strategic defense would, in the event of a planned, accidental, or minor power attack, "save lives and limit damage."[77]

For the administration, quibbles over the exact goals and potential achievements of the SDI were moot because Soviet BMD efforts and the possibility of a Soviet technological "breakout" in strategic defense compelled the United States to pursue a comparable research program.[78] Reagan spoke of massive Soviet defensive efforts and invoked the image of a "Red Shield." The omnipresent references to the Soviet BMD threat contrasted sharply with other statements by Reagan and Weinberger that parallel Soviet efforts in strategic defense should be encouraged, even to the point of America sharing its technology with the Soviets.[79] During his second debate with Walter Mondale in 1984, Reagan said that once the United States had developed the technology, then we should say to the Soviets, "Here's what we can do, we'll even give it to you, now will you sit down with us and once and for all get rid . . . of these nuclear weapons?"[80]

The administration added the ubiquitous argument that, like all offensive weapons programs, a strategic defense program would

---

Weapons, p. 13. A White House publication entitled *The President's Strategic Defense Initiative* and dated January 1985 skirts the notion of a comprehensive system completely and concentrates almost exclusively on the enhancement of deterrence through strategic defense. For a review of official goals that begin with and emphasize deterrence see Bruce, MacDonald, and Tammen, *Star Wars at the Crossroads*, pp. 13–15. For more on the shift to deterrence see Leslie H. Gelb, "Star Wars Advances: The Plan vs. the Reality," *New York Times*, 15 December 1985, pp. 1 and 34.

77. Abrahamson testimony, *Strategic Defense*, p. 13. Senator Sam Nunn in 1988 announced tentative support for what he called an "accidental launch protection system."

78. *FY 1986 Annual Report*, p. 56.

79. Speaking soon after the March 1983 address, Weinberger said, "I wouldn't mind at all" if the Soviets tried to put together their own SDI; quoted in Hoffman, "Futuristic Soviet Defense," p. 5.

80. Debate transcript, *New York Times*, 22 October 1984, p. B5.

have beneficial consequences for arms control. Active pursuit of strategic defense would induce Soviet flexibility at the bargaining table. But Star Wars would not be traded away to achieve cuts. Rather, its actual deployment would induce reductions. The reality of effective defenses, White House and Pentagon officials argued, would allow both superpowers to escape the offensive arms race treadmill and "build down" their strategic arsenals.[81] Even though the president stated several times that SDI was not a bargaining chip, many in Congress premised their support for the program on the hope that it could be traded away. Others were deterred from voting against SDI by fear of weakening the president's hand at the arms control table.[82] In effect, the president used arms control negotiations as a bargaining chip with Congress to protect SDI just as much as he used SDI as a bargaining chip with the Soviet Union.

Even if the precise strategic goal of SDI proved somewhat mercurial, the spinoffs, both military and commercial, were, for many, a concrete justification for the research program. Some in the administration, industry, and the press argued that while Star Wars might not provide an adequate defense against intercontinental nuclear warfare, it would certainly furnish a host of effective technologies for conventional conflict in the areas of command and control and naval and air defense.[83] Ever in the background of the Star Wars debate is the promise, as with the civilian space program, of numerous industrial spinoffs from the technological breakthroughs that the program could produce.

The administration's proliferation of reasons for SDI enhanced elite and popular support for the program. Many media reports

81. The president listed arms control in his March 1983 speech. George Keyworth premised his whole presentation on SDI before the Foreign Relations committee on the grounds that arms control would be the biggest payoff of the program; *Strategic Defense*, pp. 7–12. Weinberger (in his *FY 1988 Annual Report*, p. 63) and many others credited the return of the Soviets to the strategic arms talks to the SDI decision and program.

82. Senator Larry Pressler, an active participant in the SDI debates in Congress, argues that congressional support for the president's negotiations was an important influence on many members. Pressler, *Star Wars*, pp. 71 and 83.

83. See Michael Brody, "The Real-World Promise of Star Wars," *Fortune*, 23 June 1986, pp. 92–98. Brody's simple thesis is a booster for the business-world beneficiaries of SDI funding, who are ever in need of more arguments for the program: "Even if a leakproof nuclear defense system is an unattainable dream, the nation's arsenal can use some of the bits and pieces," p. 92. He argues that Israel signed on to the project largely because of the potential applications for defense against tactical missiles. See also Malcolm W. Browne, "The Star Wars Spinoff," *New York Times Magazine*, 24 August 1986, pp. 19–26+.

dwelt on the apparent contradictions among the several justifications, in particular, the contrast between Reagan's astrodome defense and bolstering deterrence. Nevertheless, these varied arguments probably elicited more support than opposition. The media reports tended to focus on confrontations between Congress and the administration, during which congressional opponents of SDI would select from the mix of rationales to attack the program.[84] But the jumble of rationales itself was not typically the basis for their opposition. The reports failed to investigate how many were convinced, or provided with leeway, to support the program by being able to pick and choose from the cornucopia of arguments. Politicians could select the most politically palatable justification. Moreover, for many, there were no contradictions; the various goals were mutually reinforcing, a "seamless web" in the step-by-step development of the program.[85] While content to leave the general public with the soothing vision of technological invulnerability, the administration and its allies produced the host of other justifications to combat a scientifically astute opposition and to sell the program to a politically divided Congress. Moreover, no matter what the objection, the administration discounted it by claiming that SDI was only a research program and that no one should be against research nor pass judgment without more knowledge. Advocates were free to pick and choose among justifications, while opponents found it difficult to be totally in opposition to a research program with so many rationales. As Les Aspin, chair of the House Armed Services committee, assessed congressional attitude on SDI in 1985: "There's the feeling that there's no really big decision to make now because it's just a research program."[86]

In fact, Congress never did make any big decisions about Star Wars even though it made a substantial cut every year. From 1985 through 1989, Congress cut about $5.4 billion from administration budgets for SDI. The largest single reduction took place in 1988, when Congress cut $1.6 billion from Reagan's $5.2 billion request. The perennial legislative ritual involved substantive debate over the need for SDI, but political divisions in Congress prevented significant policy decisions on the program. Instead, congressional input

---

84. For an example of this kind of media report see Tim Carrington, "Politics of 'Star Wars' Misfire and Plan Is Hit in Hail of Budget Cuts," *Wall Street Journal*, 22 July 1986, pp. 1, 10.
85. Nolan's phrase, in *Guardians of the Arsenal*, p. 203.
86. Quoted in Gelb, "Star Wars Advances," p. 34.

on Star Wars came down to the level of the funding cut, usually justified more on fiscal than strategic grounds. As Charles Percy argued in defense of his proposed $100 million cut in the FY 1985 SDI budget, "This amendment . . . is a deficit reduction amendment. It is a vote for fiscal responsibility. It is not a referendum on 'Star Wars.'"[87] At times, it was merely a matter of finding what price the congressional market would bear. In 1985, for example, both houses voted on a series of floor amendments calling for contrasting levels of SDI funding. In the House the level had been set at $2.5 billion in committee. Ron Dellums offered $954 million (defeated 102–340). Nick Mavroules bid $1.4 billion (defeated 155–268). Then conservative Republican Jim Courter raised the stakes to $3.7 billion and lost 104–315. Norman Dicks offered $2.1 billion (defeated 195–221). The minority made one more attempt to raise the figure, this time to $2.9 billion (defeated 169–242), before the floor finally settled on the original figure of $2.5 billion.

Expanding Budgets and Enlisting Bureaucracies and Corporations

Strategic and political justifications went hand in glove with the rapid increases in the SDI budget. Congressional cuts did not prevent SDI from quickly achieving an important bureaucratic and financial status in the Pentagon. The bureaucratic institutionalization of SDI developed from the creation of the Strategic Defense Initiative Organization and the increasing importance of SDI funds to the military services. The SDIO, created as an integral part of SDI, was a unique institution in the Pentagon: an independent office responsible for direction and interservice coordination of a major research and development program. The office was to serve several closely related purposes, including providing political prominence, streamlining the R&D process, coordinating service activities, and defending SDI from bureaucratic conservatism and interservice rivalry.[88] The formation of the SDIO gave the program a special status, centrally controlled, and not the domain of any one of the services. It

---

87. Quoted in Pressler, *Star Wars*, p. 70. Senator Pressler argues that this logic prevailed during much of the debate on Star Wars funding.

88. For more on the role of the SDIO see Gerald Steinberg, "The Bureaucratization of R&D in the Strategic Defense Initiative Organization," in Steinberg, *Lost in Space*, pp. 37–52; and Katherine Magraw, *SDI: Fading Fantasy or Fait Accompli* (Washington, D.C., 1988), pp. 6–11.

immediately gave the initiative a small bureaucracy with a stake in the project's future, in advance of establishing links to the services. The SDIO also provided organizational support for publicity and public relations. For example, the office wrote every member of Congress, offering to answer any questions about the program, and held private teach-ins for members.[89] Although the SDIO never came to dominate strategic defense policy, the organization and its director, General Abrahamson, accomplished much of their bureaucratic mission. As one Pentagon consultant said of Abrahamson's leadership, "From the Reagan administration's perspective he was good in getting the program going. He had imaginative new ideas and he spread a lot of money around to a lot of places to build up a constituency for the program."[90]

Although the president's announcement and his subsequent commitment to SDI took the military services by surprise, their initial reluctance was overcome rather quickly. The rapid inflation of the SDI budget could not be ignored. By 1985 it was clear that the buildup was reaching its apex. Pending cutbacks made any new source of money that much more attractive, and the president's commitment to strategic defense was evident in the level of funding. By 1987, SDI represented nearly 10 percent of total military R&D funding, even after substantial cuts by Congress (Table 11). Consequently, the air force, navy, and army maneuvered to get their share of the Star Wars pie, whether they were enthusiastic about strategic defense or not. All developed a substantial stake in SDI. If Star Wars were deployed, they would reap the unparalleled financial benefits. If it were not, they had lost nothing as a result. In fact, many SDI-funded programs in each service could bear fruit even if a strategic defense system were never deployed. SDI money was available for a wide variety of projects with potentially broad application, such as command and control systems and magnetic rail-guns. As a result, not only could each service play a role in the various phases of production and deployment, but each could gain immediately from the infusion of research money for projects with potential in several areas of weapons research and applications that would survive the demise of Star Wars.

89. Maxwell Glen, "'Star Wars' Future Remains Uncertain Despite Early Successes in Congress," *National Journal*, 10 August 1985, p. 1833.
90. Michael R. Gordon, "General Quitting as Project Chief for Missile Shield," *New York Times*, 28 September 1988, p. 21. See Nolan, *Guardians of the Arsenal*, chap. 5, on the role of Abrahamson and SDIO in institutionalizing the program.

Enlisting corporate support was not so simple. Defense contractors were initially wary of signing on without evidence of the administration's resolution. Many needed persuasion. Congressional commitment appeared tenuous, and some saw SDI as a program that could be retired with Reagan in 1988. The SDIO therefore launched a comprehensive pro-SDI campaign, aimed at the media, localities, and Congress. To promote the program, especially in areas not affected by major contracts to companies like Boeing and TRW, the SDIO sponsored a public relations effort aimed specifically at business and community leaders.[91] Promotional programs aside, SDI budgets reaffirmed the administration's commitment to a de facto industrial policy based on military spending. The money from the Star Wars programs constituted a large infusion of capital for advanced technology research by industries and universities, and at a time when the rest of the defense boon in high-tech R&D and procurement programs threatened to slacken. For both large and small producers in the aerospace, computer, and electronics industries, Star Wars represented potential long-term growth in a defense market that was projected to level off in the next several years. As Wolfgang Demisch, an analyst for First Boston, argued, "SDI is the future of the defense industry. No competitive high-tech company can afford not to be a part of SDI."[92] Should development and deployment of either a limited or a full-scale system occur, the seventy-five billion to more than one trillion–dollar program would be the largest military bonanza in history.[93] But just as with the military services, even if partial deployment did not occur, the contractors would benefit from many pieces of SDI that could be used for other military programs. Again, it was risky to ignore 10 percent of the Pentagon R&D budget. In fact, SDI had little trouble attracting support. In the first two years, the SDIO received ten times the number of research proposals it could fund.[94]

SDI contracts were spread thickly and distributed widely. By April 1987, the SDIO had made over thirty-three hundred contracts worth $10.9 billion.[95] Giants of the defense industry, including Lockheed, Boeing, TRW, and Rockwell, received the largest contracts,

---

91. For details see William F. Vandercook, "SDI Show Hits the Road," *Bulletin of the Atomic Scientists*, October 1986, pp. 16–18.
92. Quoted in "The Star Wars Sweepstake," *Time*, 7 October 1985, p. 48.
93. Bruce, MacDonald, and Tammen, *Star Wars at the Crossroads*, pp. 69–71.
94. Browne, "Star Wars Spinoff," p. 24.
95. Nolan, *Guardians of the Arsenal*, p. 200.

which were often substantial but not a large portion of their overall military business. Smaller firms received less substantial contracts that often represented their largest share of defense contracting. The SDIO reserved 10 percent of its budget for Small Business Innovative Research grants, and most SDI contracts were under the $2 million mark. One study concluded that SDI contracting was paralleling that of the B-1 bomber: a wide distribution of contracts, especially to the states and districts of the most influential senators and representatives.[96] The money was also available for a broad variety of research projects, and there explicitly was no preferred approach, at least initially. Futhermore, those who received SDI money were being encouraged to develop private patents and applications. Unlike past administrations, which kept tight control over technologies developed with Pentagon funds, the Reagan administration encouraged Pentagon-funded scientists to sell their research for private gain.[97] Estimates of private sales from the commercialization of Star Wars technologies are well above a trillion dollars worth.[98]

Arms Control and Early Deployment

The Reagan administration made two other interrelated commitments in its attempt to institutionalize the strategic defense program: not to let arms control impede Star Wars and to push for early deployment. At Reykjavik, Iceland, in October 1986, arms control discussions ended at a summit between Reagan and Gorbachev when the administration refused limits on strategic defense programs as a condition for an agreement on strategic offensive arms. This move compelled many in Congress to support the president, as the issue was transformed from one about expenditures into one of national political will. It also met with favorable public reaction; opinion polls showed heavy support for Reagan's actions.[99] To the Pentagon agencies and the business community, Reykjavik was an

---

96. William D. Hartung et al., *The Strategic Defense Initiative* (New York, 1985), p. 181.

97. On Reagan's private-gain policies see William J. Broad, "Space Arms Scientists in U.S. Selling Rights to Discoveries," *New York Times*, 11 November 1985, pp. 1, 12.

98. Browne, "Star Wars Spinoff," p. 22.

99. *New York Times*/CBS News polls after the summit showed 44 percent blaming Gorbachev for the failure to reach agreement (versus 17 percent blaming Reagan). Furthermore, 68 percent (versus 20 percent) thought the president should not give up Star Wars for a "big reduction" in U.S.-USSR nuclear arsenals. *New York Times*, 16 October 1986, p. 11.

assurance that the administration was committed to following strategic defense through to deployment.

Just as the administration would not swap SDI for Soviet agreement on an offensive nuclear weapons treaty, neither did it want the ABM treaty of 1972 to be an obstacle to the testing and deployment of strategic defense systems. Controversy began with the administration announcement in late 1985 that it no longer adhered to the accepted interpretation of the ABM treaty, which prohibited testing of ABM systems or components based on new physical principles. This prohibition effectively would have eliminated SDI testing. The administration argued against this "narrow" interpretation, and for a broad one that would allow testing of new systems.[100] The immediate uproar in Congress and the press over what many saw as an attempt to rewrite history compelled the administration to reach a compromise. Reagan promised to adhere to the restrictive interpretation, but said he was legally justified in moving to the broad interpretation when desirable. The dispute simmered for a year and then boiled over in early 1987, by which time the administration had begun pushing for more and earlier tests that moved the program ever closer to violating the traditionally understood treaty proscriptions. Talk of early deployment had also begun. Assuming the chair of the Senate Armed Services committee, Sam Nunn issued a report that contradicted the administration's interpretation. The Senate Foreign Relations committee released in September its own report, which went one step further. It argued that the administration's expansive view of executive treaty interpretation endangered the Senate's ability to consider and ratify the pending Intermediate Nuclear Forces Treaty.[101] Several attempts were made to restrict SDI testing to a strict interpretation via funding stipulations. Push never came to shove, however. Although SDI continued to be an obstacle in negotiations with the Soviets, testing under Reagan never breached the traditional interpretation.[102] Nevertheless, the refusal to sacrifice or

---

100. For a thorough background on the controversy see U.S. House, Committee on Foreign Affairs, Hearing before the Subcommittee on Arms Control, International Security, and Science, *ABM Treaty Interpretation Dispute*, 98th Cong., 2d sess., 22 October 1985.

101. R. Jeffrey Smith, "Foreign Relations Panel Denounces Reinterpreting of ABM Treaty," *Washington Post*, 21 September 1987, p. 10. On the fight between Congress and the administration during 1987 see Nolan, *Guardians of the Arsenal*, pp. 222–26.

102. Before the June 1988 Moscow summit, Reagan reiterated his commitment to a broad interpretation and refused a Soviet proposal for both sides to agree to a ten-year adherence to the traditional interpretation. R. Jeffrey Smith, "Reagan Drops Summit Compromise on SDI Testing," *Washington Post*, 22 January 1988, pp. 1, 13.

even compromise SDI at the arms control table and the willingness to test and strain congressional tolerance on the ABM treaty demonstrated the administration's commitment to SDI.

The controversy over the ABM treaty interpretation was precipitated by the administration's plans for an early deployment, with the installment of the first level of a phased system by the middle 1990s, a possibility suggested in 1984 by the Fletcher study talk of an intermediate option. Instead of waiting until all necessary technologies had been perfected, the Pentagon would assemble a comprehensive defensive system in stages, with initial stages being deployed as others were still being developed. Each stage would add another layer of effectiveness. Phased deployment took on political importance when, in 1987, the administration and the Pentagon began to advocate early deployment of what was coming to be known as Phase I.

Early deployment policies had the effect of accelerating testing and weapons development. This in turn boosted technical progress in support of the program, which could be used to garner more support in Congress. Several policy decisions were made to facilitate early deployment. One of Weinberger's last acts in office was to create a special office to coordinate operational testing of SDI, with the intention of expediting testing for the initial deployment system and circumventing some of the bureaucracy that confronts other programs.[103] One consequence of the push toward early deployment was a shift in research priorities. During the first few years of SDI, directed-energy weapons, such as lasers, received most of the funding, while less exotic kinetic-energy weapons lagged behind. Early deployment, however, required that the currently most effective technology be readied, which resulted in kinetic weapons receiving a greater portion of the SDI budget relative to directed-energy technologies.[104]

The combination of easy money and early deployment policies produced test results that became the basis for assertions of remarkable technological progress.[105] Although the bureaucracies had nei-

---

103. Thrish Gilmartin, "New Office to Coordinate SDI Operational Tests," *Defense News*, 21 December 1987, p. 1.
104. Congressional Budget Office, *Strategic Defenses*, June 1989, pp. 16–17.
105. In the cover letter from early 1988 accompanying the release of the 1982 Pentagon studies of the High Frontier proposal, General Abrahamson wrote Senator Johnston that the documents "should provide a keen awareness of the state of technology and analysis back in 1982 and the profound successes and advances enjoyed by the Strategic Defense Initiative after just four short years." He goes on to cite a series of tests as evidence.

ther found nor anticipated breakthroughs before the SDI program was launched, claims blossomed quickly thereafter and became ammunition for SDI advocates as well as elusive targets for opponents. Although the tests did not eliminate many basic criticisms of the program, they demonstrated enough progress to maintain political support. While technological progress had not provided the impetus for Star Wars, it did help from the early stages to perpetuate the program. In September 1987 the early deployment campaign bore fruit when Milestone I approval was granted to Phase I system architecture, with a Milestone II decision on system development to follow in the early 1990s. Much of the system remained undetermined, however, and deployment did not seem feasible before 1998. But by 1988, Star Wars was no longer an initiative; it had become a "classic weapons development program."[106]

However controversial it remained, SDI had come a long way in the four years of its formal existence, largely through this rapid creation and its accumulation of bureaucratic, economic, and technological momentum. The rapid evolution from vision to program is a partial departure from the more usual pattern of military program development, a process that often begins within the military services and their research divisions, with plans for a new system, justified by plausible strategic rationales, backed by various studies, and fueled by bureaucratic imperatives.[107] Contractors frequently push for a follow-on program and often have concrete ideas, which they develop with, or present to, the appropriate Pentagon office. Recent technological advances make the system seem feasible and innovative. In this pattern, a combination of forces, bureaucratic, rational planning, interest group, and technological, drive the process forward at the beginning. It is often only later that national politics in the form of domestic political competition begin to affect the pro-

106. Bruce, MacDonald, and Tammen, *Star Wars at the Crossroads*, p. 1. As Charles Bennett puts it, "SDI ceased to be an 'initiative' and became a 'program,' in 1987"; Bennett, "The Rush to Deploy SDI," *Atlantic Monthly*, April 1988, pp. 53–61. See Bennett for other details on early deployment as well, along with Congressional Budget Office, *Strategic Defenses*, for details on the technical components and characteristics of Phase I. Bruce et al., pp. 38–39 argue that these problems and the late 1990s deployment date meant that early deployment was no longer an accurate description.

107. Matthew Evangelista describes the more characteristic process using the single case of tactical nuclear weapons development, which he argues fits other examples of technological innovation in the arms race. He even argues, briefly, that SDI does not deviate significantly from the pattern. See his *Innovation and the Arms Race* (Ithaca, N.Y., 1988); and Evangelista, "Issue-Area and Foreign Policy Revisited," *International Organization* 43 (Winter 1989): 149–71.

gram. As examples, the ABM in the 1960s, the B-1 and B-2 bombers, and MX became embroiled in domestic politics only once their programs had reached the end of development and a production or deployment decision was at hand. With Star Wars, political considerations were decisive, in critical ways, at the outset while many of the usual forces were counteractive, negligible, or less important.[108] Once the initiative commenced, however, bureaucratic, economic, and technical momentum helped to institutionalize the program.

The transformation of Star Wars from vision to program also successfully frustrated the domestic political opponents of the administration. Star Wars gradually put the peace movement on the defensive. Strategic defense undermined some of the appeal of the waning freeze campaign with rhetoric about preventing nuclear war and escaping the dilemma of mutually assured destruction, all without any compliance or cooperation from the Soviets. As the president said, "Wouldn't it be better to save lives than to avenge them?" Strategic defense not only put the administration on the moral high ground; it also compelled the peace movement to divert some of its resources into a counterattack against Star Wars.[109] With the decline of the freeze and the rise of strategic defense as objects of national attention, the peace movement lost its own arms control agenda and its primary public relations vehicle. Instead it had to campaign for continued compliance with SALT II and for the salvation of the ABM treaty.[110] Star Wars also deflected critical peace movement and media

---

108. A parallel to the Star Wars pattern might be the Midgetman program, which was a product of the political compromise over the MX. A small group of moderate members of Congress demanded commitment to the missile in exchange for support of MX. This program born of political compromise was then forced on a reluctant air force.

109. Council for a Livable World, the largest PAC in the peace movement, began to target Star Wars as early as April 1984 in a legislative "action alert" to its membership. Star Wars was subsequently a major topic in several mailings, in spite of the warning to the movement by the council's president, Jerome Grossman, that devoting energy to Star Wars was playing on Reagan's turf. Moreover, the council sponsored at least one conference on SDI and printed several special pamphlets on the subject. And the council is hardly alone in this. Several other organizations started special campaigns to stop Star Wars, including Common Cause and Mobilization for Survival. A coalition of the major arms control lobbies formed the Space Policy Working Group, which met weekly to discuss anti–Star Wars strategy. See also "The Selling of Star Wars," *Nuclear Times*, May/June 1985, pp. 10–13.

110. Several mainstream peace groups devoted mailings to these causes. The administration's actions even prompted the formation of a coalition devoted exclusively to the ABM Treaty: the National Campaign to Save the ABM Treaty. Meanwhile the freeze all but disappeared from movement literature.

attention away from offensive nuclear weapons programs sought by the administration.

Star Wars not only disrupted the opposition; it also attracted support. While the direct impact of Star Wars as an influence on public opinion in the months just after the decision is difficult to trace, a solid majority of Americans did favor the development and deployment. Early in 1985, with the actual SDI program barely a year old, 52 percent of those who were aware of the program favored it, while 38 percent opposed it. Additionally, 47 percent (against 32 percent) thought development would facilitate arms control agreements, and 50 percent (against 32 percent) believed development would help prevent nuclear destruction.[111] By December 1985 the margins had increased significantly in favor of the program. For example, 61 percent of the aware group favored deployment, and 58 percent thought it would make the world safer. A consistent but shrinking majority favored continued development throughout the Reagan presidency.[112] Again, public opinion was not wholly consistent; for a majority still favored the freeze and cutting defense spending. The popular movement that had sought to advance these goals was subsiding, however, even as Reagan began his second term. At the same time, another defense policy movement was gaining momentum. A coalition in Congress, which was caught in the middle of the nuclear weapons controversy, was developing its own agenda in the battle over defense policy. In the months after the 1984 elections, the Reagan administration would confront a different attack on its military program, when the military reform movement in Congress hit full stride with its campaign for reform of the Pentagon procurement process.

---

111. The figures in this paragraph are from *Gallup Reports* 234 (March 1985): 12–16; and ibid. 243 (December 1985): pp. 3–6.

112. Daniel Yankelovich Group polls, reported in *Americans Talk Security*, nos. 10 and 11, December 1988, p. 52.

7

Congressional Procurement Reform

When Richard DeLauer, under secretary of defense for research and engineering, moaned, "They're slicing us up into little bitty pieces, creating motherhood issues around all the pieces and then getting legislation passed,"[1] and John Isaacs, legislative director for Council for a Livable World, said, "I have to be a little cynical about the caucus. . . . The aims are noble, but the primary effect for many members is providing cover. Someone who's supporting the B-1, MX and [the Star Wars program] can say, on the other hand, 'Look I'm in favor of reform. I'm looking for more efficiency and better Pentagon management,'"[2] both of them were complaining about the same group in Congress: the Military Reform Caucus. DeLauer was an architect of and an advocate for the Reagan military buildup. Isaacs was and is chief lobbyist for one of the largest public interest groups in the peace movement. Yet both were critical of the Military Reform Caucus and its cause. These antagonists in the defense policy conflict found common ground in their shared view of the reform movement as politically motivated and a threat to the defense policy agendas they represented.

---

1. Quoted in "DeLauer: 'Caucus Is Slicing Pentagon into Itty Bitty Pieces,'" *Defense Week* 30 (April 1984): 7.
2. Quoted in David C. Morrison, "Caucusing for Reform," *National Journal*, 28 June 1986, p. 1599.

The nuclear freeze proposal and Star Wars stood as diametric visions and alternatives for U.S. defense policy. Congress was the principal battle arena for these polar forces, the vortex of the storm over defense policy. Neither alternative was politically palatable to many in the fragmented legislature, which was having enough trouble coping with the Reagan revolution. Consequently, Congress did more than react to the actions and initiatives of the peace movement and the administration; it developed its own military policy agenda. A large group in Congress developed military reform as a response to both the Reagan buildup and the peace movement. As the peace movement lost momentum during 1984, the congressional reform efforts emerged as the next significant challenge to the Reagan military buildup.

Specifically, the military reform coalition in Congress initiated a wide variety of attempts to reform the military procurement process. Procurement reform is not a single initiative like the nuclear freeze or Star Wars, but that is not surprising given the nature of Congress and the reform movement it spawned. Befitting an ideologically diverse and individualistic institution, procurement reform became an umbrella for numerous investigations and laws sponsored by diverse combinations of senators and representatives. Despite the legislative success of the military reform agenda, reformers could not overcome both internal congressional political fragmentation and external resistance from the administration and the Pentagon. While the reform coalition in Congress sustained an unprecedented assault on the conventional administration of Pentagon programs, especially in procurement, the Reagan administration and its congressional allies were able to obstruct and frustrate many reform efforts.

Congressional Reform: Eliminating Waste, Fraud, and Abuse in the Pentagon

It was only natural that as the political pitfalls of the strident debate over nuclear weapons and strategy became apparent, members of Congress would search for a different approach to national security. The publicity created by the spare parts scandals provided a politically powerful focus for reform activities: exposing and correcting waste, fraud, and abuse in the Pentagon. The congressional military reform effort soon developed a constituency and an agenda that were to make it an influential force in national politics.

The Military Reform Caucus, the institutional spearhead of the congressional reform effort, led by Gary Hart and William Whitehurst, enlisted about 50 members over the course of 1981 and 1982. From 1983 through 1984, the caucus was chaired by the all-Republican team of Representative Jim Courter and Senator Nancy Kassebaum. Membership grew more slowly, to about 80. The next leadership was expressly bipartisan, with a liberal and conservative from each chamber, namely, senators Charles Grassley, an Iowa Republican, and Arkansas Democrat David Pryor, along with representatives Denny Smith of Oregon and California Democrat Mel Levine. From the beginning of 1985 through 1986, membership jumped to more than 130, including more than 100 representatives and nearly 30 senators, almost evenly divided between Republicans and Democrats and spanning the political spectrum. By 1986 its size and ideological diversity were greater than any other nonregional congressional caucus.[3] Although the purely Republican leadership of Courter and Kassebaum may have retarded growth, the four-person bipartisan leadership and the subsequent growth in membership testify to the ever-increasing popularity of the issue, especially after the 1984 elections.

The political appeal of military reform also broadened as the focus of military reform shifted. Although the MRC had begun as an educational discussion group, the political situation came to demand action, and the potent effect of the spare parts scandal pointed the way. Legislation began to attack many aspects of the procurement system, including a few specific weapons, leading to DeLauer's not unique charge that the movement was carving up the defense debate into little "motherhood issues." Thus the movement's popularity grew with its increasing influence and its changing focus. The two went together. That is, as the MRC moved from doctrine to specific weapons to issues of waste in procurement, its constituency grew as more and more senators and representatives sought solutions to their defense policy dilemmas. As the constituency and agenda came together, the military reform movement's legislative efforts grew steadily until they reached a peak in 1985 and 1986. In fact, 1985 was the busiest year for reform legislation and also the most successful in terms of margins of victory for reform amendments and laws.[4]

---

3. For example, the MRC was only a bit smaller than the combined Northeast-Midwest coalitions. It was about the same size, but more diverse, than the liberal Arms Control and Foreign Policy Caucus.

4. This conclusion is based on a survey from 1981 to 1987 of all floor votes taken in

The shift to procurement reform began in 1983, when the leadership of the caucus passed to Courter and Kassebaum. Under their leadership, the emphasis changed from discussions of strategy and miscellaneous amendments on specific weapons to broader solutions for procurement problems. Courter wanted the caucus to move beyond discussion because he "thought it would not be taken seriously by anybody unless it became a player."[5] Gradually, much of the caucus, and other sympathetic members of Congress, found that they could often agree on broad procedural reforms of Pentagon operations such as procurement instead of on direct legislation on specific programs. One of the earliest successful MRC initiatives was 1983 legislation, compelling the Pentagon to create and staff an independent office of operational testing and evaluation. Kassebaum said that the caucus discussed the situation "and decided to move more away from the theoretical discussions and to pick up some issues we felt we could all support. We were fortunate in having an independent operational test issue which we were successful in getting behind."[6]

Fly Before You Buy: Operational Testing and Sergeant York

Troubled and troubling weapons programs, including the M-1 tank and the Maverick missile, imparted renewed relevance to a long-standing aphorism of reformers: Fly Before You Buy. That is, rigorous testing of any weapon, preferably under battlefield conditions, should precede any procurement decision. The Office of Operational Testing and Evaluation law was to assure that weapons would undergo thorough and realistic tests prior to procurement decisions. The legislation made the director of the office the senior official on operational testing and evaluation in the Pentagon. The director was to prescribe all testing and evaluation procedures, monitor all testing and evaluation, and analyze the results of tests for all

---

both the House and the Senate. Floor votes are one measure of an issue's importance and popularity. A more comprehensive survey of committee work would have to incorporate several committees and subcommittees because an important aspect of military reform was that it took place largely outside of the usual defense power structures such as the Armed Services committees.

5. Courter is quoted in Morrison, "Caucusing for Reform," p. 1597.

6. Quoted in ibid. Dina Rasor recounts the origins of the law, stating that she gave the idea to Senator Pryor and had her sources help Pryor's office directly in the drafting of the bill. Rasor, *The Pentagon Underground* (New York, 1985), pp. 275-76.

major weapons programs. The measure passed both houses by substantial margins in the summer of 1983.[7]

The law was not implemented, however, for eighteen months. The Defense Department did not request funding for the office during the next fiscal year. Weinberger failed to comply with the measure by not nominating a permanent director.[8] Some members of the MRC favored an air force colonel who had been critical of several programs in his work in the Pentagon's developmental testing office. The Pentagon was at that very time planning to eliminate this colonel's job and transfer him to work unrelated to testing. Instead, in 1985 the president finally nominated John E. Krings, a twenty-five-year employee of McDonnell-Douglas, where he had been a test pilot. The leadership of the caucus opposed the nomination, arguing that Krings's background did not imply the independence required for the job. The nomination passed by a vote of 73–18, with the negative votes coming mainly from caucus members, especially Democrats. Not only did the administration delay implementation, then, but it also staffed the office with someone unacceptable to the principal advocates of the law.

Meanwhile, the reform triangle had exposed a particularly egregious example of how the procurement process can go wrong. The fight against the Sergeant York, or DIVAD, gun tested the power of the reform movement and the integrity of their newly created Office of Operational Testing. The DIVAD program had begun in the 1970s with the mission of providing comprehensive air defense for ground forces in combat, especially for protecting tanks from air attack. This entailed the ability to track and destroy both low-flying jets and helicopters. The Sergeant York, with a fully automated tracking and firing system, was to accomplish this. It was composed of twin 40-mm guns mounted on a tank chassis and aimed by a radar adapted from the F-16 aircraft. After development work in the late 1970s and 1980, the aerospace division of the Ford Motor Company was awarded a production contract in 1981, with a programmed goal of procuring 618 guns at a total cost of $4.5 billion.

A year later the *Atlantic Monthly* published an article by Gregg Easterbrook in which he chronicled the problems that had plagued

---

7. The act was passed as an amendment to the FY 1984 authorization act for the Department of Defense and became part of Public Law 98-94.

8. On the Pentagon reluctance to implement the legislation see Michael R. Gordon, "Help Wanted in Weapons Testing Office But Pentagon Slow to Fill Top Job," *National Journal*, 14 October 1984, pp. 1914–17.

the DIVAD program.⁹ The comprehensive defense mission had been gradually downgraded as it became apparent that the gun could not handle moving targets unless they moved in predictable ways. The army had dropped jets from the defense mission and emphasized hovering helicopters, which the gun had also often failed to hit.¹⁰ Ford had worked with Department of Defense (DOD) officials to put the best face on poor test results before the production award. Finally, Easterbrook charged that Ford had been awarded the contract over its rival, General Dynamics, partly as an economic bailout.

Later, Easterbrook wrote that no major media pursued the story because it was based largely on off-the-record sources.¹¹ Then Dina Rasor's Project on Military Procurement took up the effort, especially after one of Easterbrook's sources leaked documentation to the group. This, as Easterbrook notes, gave the mainstream press hard data and an organization to blame if trouble over the story arose. Soon press reports appeared in the *Los Angeles Times* and the *New York Times*. Gradually evidence of the system's problems accumulated.¹² The GAO had done critical studies, dating back to 1980, which the Pentagon had ignored. In July 1983 the Pentagon's inspector general began an investigation that would show that the army withheld important test data from the contracting-decision process in 1981. Soon congressional hearings began in the Senate Armed Services and Governmental Operations committees.

Initially, however, congressional concern was limited. Larry Smith, a first-term Florida Democrat, was somewhat of a maverick on the issue. In May 1983 he sponsored an unsuccessful amendment to the FY 1984 Defense Authorization Bill to cut $671 million for DIVAD procurement. The measure lost by a considerable margin, 134–283. The next May, he sponsored a measure to bar expenditure

---

9. Easterbrook, "DIVAD," *Atlantic Monthly*, October 1982, pp. 29–39.
10. Easterbrook relates what became one of the most repeated DIVAD anecdotes: During a test in 1981, instead of firing at a hovering target helicopter, the DIVAD's computer-controlled turret swung toward a reviewing stand filled with military officers. The gun was mechanically prevented from firing at the reviewing stand, but in subsequent tries it also failed to hit the helicopter.
11. Easterbrook relates the development of the DIVAD story and the Project on Military Procurement's involvement in his "Why DIVAD Wouldn't Die," *Washington Monthly*, November 1984, pp. 10–22.
12. It was commonly alleged that some disclosures were made by missile advocates or allies of General Dynamics within the military who had lost in the original DIVAD decision.

on DIVAD procurement pending further test results. It too was defeated, this time by a vote of 157–229, providing another indication that military reform did not catch on until after the 1984 elections.

Representative Denny Smith, cochair of the MRC, became a leader in the fight against DIVAD in 1984. With some help from whistleblowers and a changing political climate, the Oregonian succeeded where his Floridian colleague had fallen short. Using inside information, Smith continued to expose what he and others charged were rigged tests. It was patently obvious that DIVAD had been unable to perform adequately even its least taxing mission, hitting hovering helicopters. Moreover, investigations led to accusations that Ford had overcharged for the program. D-day in the struggle against DIVAD occurred during the FY 1985 budget hearings in September of 1984. The uproar had attained front-page status in newspapers. Several congressional committees had pursued the issue. Finally, the House and Senate voted to limit appropriations and bar further procurement pending new tests. Weinberger decided to delay further procurement appropriations and scheduled a special series of operational evaluation tests. By this time the press felt free to ridicule the gun that couldn't shoot straight.[13] Even *People Magazine* did a story on the gun and Smith's role in exposing its flaws. Despite the halt in appropriations, $1.5 billion had already been spent or appropriated for the first 146 DIVADs, and about 21 had already been built. In the interim between the cutoff and the tests scheduled for the summer of 1985, the army looked for solutions. It went so far as to float the idea that it would add Stinger missiles to the DIVAD, perhaps in an attempt to placate missile advocates and other critics. The idea died quickly, especially since Stinger was hardly a trouble-free program. The army also began to investigate wholly new alternative programs; DIVAD's demise was in the wind.

Krings became director of Operational Testing and Evaluation (OT&E) just in time to confront the DIVAD imbroglio, and reformers kept a watchful eye on his handling of the problem. The summer 1985 tests showed only "marginal improvements" in the system,[14] and in late August, Weinberger killed the program. About $1.8 billion had been spent to develop and produce sixty-five DIVADs,

---

13. Including *U.S. News and World Report*'s scathing story on Sergeant York: "Anatomy of a Pentagon Horror Story," 15 October 1984, p. 69.

14. Weinberger's phrase, quoted in Bill Keller, "Pentagon Cancels Antiaircraft Gun 'Not Worth Cost,'" *New York Times*, 27 August 1985, pp. 1 and 18.

which, following cancellation, were to be used for experiments and spare parts. This was the only major weapons program terminated by the Reagan Pentagon. Weinberger had little to lose by ending DIVAD. The mission of division air defense would still get a replacement program. A little more than two years after DIVAD was canceled, the army announced a new winner in the search for an air defense system: ADATS, which stands for air defense antitank system. Furthermore, as some congressmen noted, Weinberger had little credibility on Capitol Hill because of his sanguine intransigence on defense spending. The cancellation boosted his credibility among some members.[15] Krings's performance, credibility, and the potential of the OT&E office, received mixed reviews. Some credited his report on the final testing results as an influence on Weinberger's decision. Others claimed Krings originally wrote a mildly critical report, which was rewritten after another Pentagon agency group wrote a scathing report on DIVAD's performance.[16] The effectiveness of the office was a matter of dispute from then on. In 1987, the GAO issued a critical evaluation of the office's work. The staff did very little field work at test sites and relied predominantly on reports, not even raw data, submitted by the military service doing the test.[17] Members of the MRC charged that unrealistic tests were being rubber-stamped by Krings's office, and they made more than one unsuccessful call for his resignation.[18]

The DIVAD case illustrates the strengths and weaknesses of the military reform movement. First, the reform triangle proved its potency; for DIVAD was the first major weapons program in production since the 1960s to be canceled. Yet it was also the exception that proved the rule; for it was the only system thus stopped. Other extremely problematic programs, both big and small, hobbled on in spite of equally aggressive attacks. These include the Bradley Fighting Vehicle, the Abrams M-1 tank, and the Maverick antitank mis-

---

15. See comments of senators Pryor and Quayle and Representative Smith in George C. Wilson, "Weinberger Kills Antiaircraft Gun," *Washington Post*, 28 August 1985, pp. A1, 14.

16. Critics also charged that Krings's office did not do enough to stop the problematic Advanced Medium-Range Air-to-Air Missile (AMRAAM) program. See David C. Morrison, "Trying Times for Weapons Tester," *National Journal*, 19 April 1986, pp. 946–47.

17. Fred Kaplan, "Watchdog's Report Lacks Bite," *Washington Post*, 27 March 1987, p. 21.

18. Krings remained director until late in 1989, when the Bush-Cheney appointee, Robert C. Duncan, was confirmed by the Senate.

sile. In some ways, DIVAD was a sacrifice meant to appease angry gods on Capitol Hill. But one weapon was an insufficient offering.

More Bang for the Buck: Legislating Clean and Competitive Procurement

As military reformers were dismantling the Sergeant York program, they were assembling a battery of proposed procurement reform legislation. Just as Congress became the institutional base of reform, procurement reform became the staple reform issue. Despite problems with the OT&E office in practice, most reformers believed that sweeping procedural reform of the procurement process was the only solution. However necessary, attacks on individual weapons, such as the Sergeant York, were time consuming and aimed at the wrong end of the process. Reform needed to correct things at the start, before problems appeared. Procedural reform also avoided dividing the MRC over specifics. This approach was attractive both to sincere reformers who felt this was the only way to rectify problems in the Pentagon and to those who needed a way to confront the worst excesses of the buildup without getting caught in the politically treacherous debates over individual weapons, overall budgetary priorities, and military strategy.

Perhaps the principal axiom among procurement reformers was that the Pentagon needed a dose of free-market discipline. The tongue-in-cheek motto of the movement was More Bang for the Buck. The fundamental goal was efficiency, and the greatest promoter of efficiency, from the reformers' perspective, is free-market competition. Hence much of military reform legislation aimed at increasing the level of private competition for defense contracts, an aim that drew considerable resistance from the administration. The first such piece of legislation was the Competition in Contracting Act (CICA) of 1984, which was "widely regarded as the first major piece of procurement reform legislation passed by the Congress in over 40 years."[19] Although the act was directed at the entire federal procurement program, the problems posed by the Pentagon's disproportionate share of that program provided the major impetus for the legislation. The authors of the original version of the act in 1983 were

---

19. Opening remarks by Representative Jack Brooks, in U.S. House, Committee on Government Operations, Subcommittee on Legislation and National Security, *Constitutionality of GAO's Bid Protest Function*, 99th Cong., 1st sess., 28 February and 7 March 1985, p. 1.

senators William Cohen and Carl Levin, junior members of the Armed Services committee and of the Governmental Affairs committee, which also had to approve the bill. During Armed Service committee hearings on the act in 1983, committee chair John Tower, hardly an advocate of the measure, opened the hearings by opining that "nothing compromises national confidence in its military capabilities more than the perception that defense procurement procedures are inefficient or wasteful."[20]

The goal of CICA is to promote "full and open competition" in federal procurement procedures through a number of procedures and requirements.[21] Most of CICA's new or amended procedures are intended to alter or eliminate practices that limit competition in the awarding of procurement contracts, such as failure to advertise the potential contract and writing advertisements so as to exclude all but one or two potential contractors. The act requires all federal agencies to open bidding on contracts to all qualified and interested firms and then enforces this requirement. It prescribes appeals procedures for competitors who feel they have been denied the opportunity to compete: the protest is taken to the GAO, which reviews and decides the case. The act authorizes the GAO to suspend or stay contract actions during the course of the bid protest process and gives the agency the power to award costs and fees to a protester who wins.

The bill passed as part of the Gramm-Rudman Deficit Reduction Act of 1984. Immediately Reagan put Congress on notice that he considered the bid protest function of the GAO, specifically the provisions for contract suspension and the awarding of costs and fees, unconstitutional. The administration argued that the act gave the comptroller general, the head of the GAO, executive powers—the power to direct executive agencies. The GAO and comptroller general, however, are not part of the executive branch; they are a creation of Congress to help the legislature fulfill legislative functions. To the administration, the power of the comptroller general to direct executive agencies was equivalent to the legislative veto the Supreme Court had ruled unconstitutional the year before.[22] The administration subsequently ordered all agencies not to comply with the bid protest provisions, which were the teeth of the act. This

---

20. U.S. Senate, Committee on Armed Services, *Competition in Contracting Act of 1983*, 98th Cong., 1st sess., 7 and 8 June 1982, p. 43.

21. See Andrew Mayer, *The Competition in Contracting Act*, Congressional Research Service, Report No. 85-115F, 14 May 1985.

22. Immigration and Naturalization Service v. Chadha (1983).

nullification led to a little-noticed struggle between the branches. Hearings were held to investigate what some in Congress charged was the executive branch's unlawful usurpation of judicial power, and the failure of agencies to carry out the law.[23] Congressional committees threatened to withhold funds for the Justice Department if it did not order agencies to comply with the act. A federal court eventually ruled against the administration. Attorney General Meese relented, but the administration also appealed the case.[24] The next year, the bid protest procedures were cast back into constitutional limbo when the Supreme Court ruled unconstitutional the part of the Gramm-Rudman Deficit Reduction Act that gave the comptroller general executive powers.[25]

While the constitutionality of CICA was being contested, military reformers continued efforts to force competition and other procurement reforms on the Pentagon. In 1985, during the legislative work on the FY 1986 defense authorization bill, reformers championed a long list of reform measures that culminated in the Defense Procurement Improvement Act of 1985. This act, together with additional work in 1986 on the FY 1987 defense authorization bill, became the high point of reform legislation efforts.

The extensive list of procurement reform proposals that dominated congressional action on the defense authorization bill during the summer of 1985 offers a convenient summary of the range of reform legislation. Heading the reform agenda were more stringent requirements for competitive contracting, specifically, "dual sourcing" and "creeping capitalism." The dual-sourcing amendment required the Pentagon, except under certain circumstances, to seek multiple contractors for both the development and production stages of procurement. Under creeping capitalism the portion of competitive contracts was to increase by 5 percent each year until the total percentage reached 70 percent. Moreover, contracts would be renegotiated on a yearly basis. "Creeping capitalism" was a particular interest of Senator Charles Grassley, a cochair of the MRC at

---

23. U.S. House Committee on Government Operations, Subcommittee on Legislation and National Security, *Constitutionality of GAO's Bid Protest Function*. These hearings contain testimony by the comptroller general, Justice Department officials, and David Stockman and include relevant documents which detail the arguments for and against the constitutionality of the bid protest procedures.

24. The case is Ameron Inc. v. U.S. Army Corp of Engineers, (D.N.J. 1985). See also Myron Struck, "Meese Averts Showdown on GAO Contract Power," *Washington Post*, 5 June 1985, p. 21.

25. Bowsher v. Synar (1986).

the time. Another amendment, the Boxer-Bennett amendment, confronted the "revolving-door" problem of personnel transfers from the Pentagon to defense industries and vice versa. In particular, the steady flow of Pentagon employees from government jobs to employment with defense contractors blurred the distinction between public responsibility and private interest and, many thought, contributed to waste and fraud. The problem was especially evident in the significant numbers of Pentagon officials who left the government to work for the very contractor and even the particular program over which they had supervisory and decision-making responsibility. The Boxer-Bennett amendment called for strictly supervised limitations on revolving-door employment; pentagon employees would be prohibited for two years after leaving government employment from working for contractors over whom they had significant responsibility. The legislation also sought to enhance and enforce reporting procedures and penalties for violations.

Under the rubric of "allowable costs," yet another amendment forbid the Pentagon to reimburse costs incurred by contractors for such things as lobbying, legal fees, advertising, entertainment expenses, and other amenities that often found their way onto bills to be paid by Uncle Sam. Another compelled the Defense Department to utilize "should-cost studies," whereby contractors would be required to submit data on how their projected costs of production compare with similar operations performed in civilian production. Other proposed legislation required warranties on Pentagon purchases. All but one of these measures was adopted, in some form, by Congress in 1985, and MRC members sponsored all of them. But many were passed in diluted form, and the losses came at the hands of conservatives, especially under the leadership of Republicans in the Senate Armed Services committee. The battles in the summer of 1985 revealed that while nearly everyone in Congress could claim to be in favor of reform, agreement in principle did not always mean agreement on substance, even among self-described reformers.

Both the powerful appeal and the limitations of the reform agenda are evident in voting behavior on the House and Senate floors. Along with the rest of their colleagues, the caucus membership divided sharply along partisan and ideological lines over most issues of nuclear weapons, strategy, and arms control. On an amendment to bar testing of ASATs, which passed by a vote of 229–193 despite presidential opposition, the House caucus membership split 60–37, largely along party lines, but with several liberal Republicans voting for the amendment. On a proposal to slash SDI funding from $2.5 billion to $1.4 billion, rejected 155–268, the caucus members mirrored their colleagues by dividing 35–59, with southern Democrats providing the margin for the nays. Similar patterns emerged from the series of seven floor votes taken on the MX missile over the course of the year.

On reform amendments, a different pattern emerged, for both the membership as a whole and the caucus in particular. The caucus would frequently unite on reform measures, but then again, so would the whole House or Senate. Out of eleven distinct votes on reform measures in 1985, the reform position won nine times, with an average gap between ayes and nays of 309 votes. This contrasts sharply with any set of votes on nuclear weapons or arms control issues. For example, on the seven MX votes, the average gap between ayes and nays was only 24 votes.[26] What this indicates is the extent to which procedural or structural reform became politically appealing and often irresistible.

Many reform measures were opposed, however, and victories often required compromise. The MRC was tested by substitute amendments designed to dilute or kill tough reform measures, and these substitutes were sometimes offered by conservative MRC members who were also on the Armed Services committee. For example, the Boxer-Bennett revolving-door restrictions passed by a 397–19 vote after the defeat of a weaker version offered by conservative Democrat John Spratt, a member of the Armed Services committee and the caucus. Other measures passed by equally large margins after compromises made before the final vote. For example, Jim Courter's creeping-capitalism amendment mandated 40 percent competitive procurement in FY 1986 and 5 percent annual increases thereafter until 70 percent of Pentagon procurement was contracted competitively. Bill Nichols of the Armed Services committee altered Courter's proposal by changing the percentage increases from requirements to goals, and the amended proposal passed by a 416–0 vote. Changing places, Courter, a former cochair of the caucus and a Republican on the Armed Services committee, proposed his own killer amendment to Barbara Boxer's should-cost legislation. The Courter substitute was defeated 189–232, with the caucus splitting 37–61. The Boxer amendment was then passed 384–31, with the

---

26. The gap on two votes on chemical weapons was 30, and on the single vote on an ASAT testing ban, the gap was 36.

caucus voting 91–4. While the vote on the Courter substitute shows that caucus membership helped to defeat the measure, in most cases of controversial measures, the caucus divided along partisan lines closely paralleling the floor vote as a whole.[27]

Occasionally, however, a procedural reform proved too controversial. Democrat and caucus member Dennis Hertel offered a tough amendment that would have empowered the inspector general of the Defense Department to suspend contract payments and debar contractors if he found waste, fraud, and abuse in connection with the contract. It was defeated 176–240, with a solid Republican bloc aided by 60 votes from southern Democrats. The amendment split the caucus right down the middle, 48–49, with Republican members voting 9–39 and Democratic members voting 39–10.

Similar, although often less stringent, amendments were sponsored and passed on the Senate floor, with largely analogous voting patterns. For example, an amendment sponsored by senators Pryor and Grassley required the Pentagon to keep two sources competing for each major weapons program from development through procurement. Only under particular circumstance could this be waived; in fact, for major systems, a waiver required legislative approval. This measure lost, by a vote of 22–67, to a watered-down alternative sponsored by the Armed Services committee, under the leadership of Dan Quayle, then an obscure first-term senator, who was chair of the Armed Services Defense Acquisition Policy subcommittee, created in 1985 to handle the tide of reform legislation for the Armed Services committee. On the 22–67 vote, Charles Grassley, a reform stalwart, was the only Republican in support of the measure. Overall, the caucus split 9–18 against the measure, with Democrats divided 8–8. When the Armed Services committee offered its diluted alternative to Pryor's package, the MRC decided to support it. It passed 89–0. The Quayle substitute made it much easier for the Pentagon to avoid the dual-source stipulation. The Senate version of revolving-door legislation significantly narrowed the number of officials covered by the regulations and had no two-year prohibition against taking a job with a contractor after leaving government employment.[28]

---

27. For example, on an amendment to alter the Boxer should-cost legislation slightly, 58 percent of Democrats voted against the amendment as did 65 percent of MRC Democrats. The voting by MRC Republicans even more closely matched the overall GOP vote.

28. In September 1988 after Quayle unexpectedly became George Bush's running mate, Mel Levine, a liberal member of the MRC, used Quayle's performance in the

Congressional Procurement Reform

As these examples indicate, much of the opposition to reform measures came from the Armed Services committees in both houses and what many perceived as jealous protection of legislative turf. On several of these votes, members of the Armed Services committees composed a disproportionately large share of the opposition. For example, on the should-cost vote, twelve of the thirty-one nays came from Armed Service members. As Representative John Bryant said of the committee's opposition during the 1985 floor action, "What you're implying today is, 'we don't want any outsiders to be part of the process.'"[29]

The struggle over what kind of reform package would emerge took place in the House-Senate conference.[30] Indeed, military reform dominated the conference, even with such controversial issues as the MX and Midgetman missiles and Star Wars on the agenda. A conference comparison of the House and Senate reform proposals filled ninety-one ledger-sized pages covering thirty-six issues. On the most important measures, including revolving-door and dual-source procurement, the more restrictive House versions were compromised; the Senate Armed Services position, represented by Dan Quayle, prevailed. Negotiations left both the revolving-door and dual-source provisions with significant exceptions and loopholes that critics charge eviscerated the laws.[31] Creeping capitalism disappeared altogether. Compromises and defeats notwithstanding, a substantial number of reform provisions were added to the authorization bill. Aggregated, they became Title IX of the 1986 Defense Authorization Act, more commonly referred to as the Defense Procurement Improvement Act of 1985.

The process did not end there, however. Reformers sought to strengthen certain measures while their opponents sought to repeal

---

1985 and 1986 reform negotiations as yet another criticism of the GOP vice-presidential candidate; see Levine, "Quayle: The Defense Industry's Champion," *New York Times*, 26 September 1988, p. 23.

29. Quoted in *Congressional Quarterly Weekly Report*, 29 June 1985, p. 1266.

30. Pat Towell, "Pentagon's Buying Practices: Battle Lines Drawn," ibid., 13 July 1985, pp. 1369–72, and Towell, "Conferees Agree on Defense, But Nerve Gas Is Sticking Point," ibid., 27 July 1985, p. 1473.

31. The revolving-door law adopted the House's two-year ban but limited it only to presidential appointees who were "primary government representatives" in negotiating the contract. Very few of the thousands involved in revolving-door employment fit this description. The dual-source compromise allowed the secretary of defense to waive a second source for reasons of "national security," and the compromise dropped the House limit to exceptions to 50 percent of procurement contracts. See Public Law 99-145, sec. 912 and sec. 921; and *Congressional Quarterly Weekly Report*, 3 August 1985, p. 1537.

or enervate some provisions of the 1985 act. This fight became part of the work on defense authorization in 1986. In particular, reformers sought to toughen up the revolving-door regulations. Led again by Dan Quayle, the opposition sought, among other things, to repeal the should-cost requirements set in motion by the 1985 act. Reformers eventually prevailed on both issues. The application of the revolving-door law was broadened and reporting requirements were specified. The should-cost requirements in fact were elaborated as well. Other reform measures not addressed in 1985 were added, too, including regulations to protect whistle-blowers and specific procedures intended to reduce the cost of spare parts. All these, along with reforms that grew out of the Packard Commission (discussed below), became the Defense Acquisition Improvement Act of 1986.[32]

The procurement reforms of 1985 and 1986 revealed the strengths and weaknesses of the reform coalition. While the ideological diversity of the reformers, as exemplified by the Military Reform Caucus, provided nonpartisan power and appeal, it also set limits on political unity—the first paradox of reform. The activist members helped organize defense reform into a powerful political issue and then drafted and promoted legislation that often attracted overwhelming majorities. But those majorities, including moderates and conservatives, often depended on the procedural flexibility of the legislation. That is, much of reform legislation involved procedural compromise rather than detailed, nonincremental policy change. Successful reform measures tended to be those that required the Pentagon to add new personnel or change a process rather than meet particular requirements or adjust to a different budget. This allowed many of the measures to be passed but also set limits on the potential impact of reform legislation—the corollary paradox of reform.

Reformers and the Defense Budget

While the net effect of all this legislative effort on the quality and efficiency of procurement proved somewhat ephemeral, as we shall see, the political effect was more immediate and palpable. Regardless of its impact on the quality of Pentagon procurement, the reform movement was helping to bring the Reagan buildup to a halt. First and foremost, military reform and the issues of waste and

---

32. The Defense Acquisition Improvement Act of 1986 is Title IV of the Defense Authorization Act of 1987.

fraud had tarnished the "every dollar is necessary' rhetoric employed by Reagan, Weinberger, and their congressional allies. Second, the political toll taken by the deficit and cuts in domestic programs made the Pentagon an attractive if not a necessary target. Consequently, some reformers were willing to venture beyond procurement reform and confront the Pentagon budget head on. In fact, without the budget crisis of the 1980s, the military reform movement itself probably would never have taken hold in Congress. Although many reform advocates were staunch allies of the Pentagon, the combination of the widespread perception of the indiscriminate nature of Reagan's spending with record deficits compelled many to seek arguments that held the line on spending while avoiding alliance with the peace movement. More and more members of Congress were finding excess to be trimmed from the defense budget regardless of what the Soviets were up to.

One attack on the Pentagon budget used the tried and true theme of waste, fraud, and abuse. The MRC and the Armed Services chair Les Aspin exposed the Pentagon's inflation bonus, whereby it calculates its own inflation factor for purposes of procurement budgeting—yet another reason the Pentagon budget was difficult to control. One of the Pentagon's own reform initiatives back in 1981 had increased the factor to avoid the funding shortfalls that had plagued past procurement efforts when inflation exceeded estimates. But actual inflation in the early 1980s had proved much lower than the revised Pentagon calculation. Given the enormous increases in the first Reagan military budgets, the exaggerated inflation adjustment translated into a significant bonus (somewhere over thirty-six billion dollars for the first four years of the Reagan administration from 1982 through 1985) that the Pentagon never refunded. Aspin and the caucus wanted the Pentagon to account for the money and return it to the treasury. Again, this effort avoided issues of priorities and simply attacked the Pentagon on the unassailable issue of waste and abuse. Although monetarily very significant, the issue lacked the visceral public impact of extremely expensive hammers, coffee pots, and toilet seats. The Congressional Budget Office reported that Congress got some of the money back by transferring excess funds from prior years to new obligations, but the percentage of the inflation bonus recovered could not be calculated.[33]

The constituency of those who sought to slow down or cut growth

---

33. This brief description does not do justice to the causes and complications of the inflation-adjustment imbroglio; see Senator David Pryor's "Defense Department's In-

in the military budget, particularly to limit deficits, was not confined to liberal Democrats. Many in the Republican congressional leadership sought to persuade the administration to slow down its buildup. Some conservatives used elaborate arguments linking deficits to future inflation, which would in turn hurt defense industries. Some southern Democrats, like G. V. Montgomery of the Armed Services committee, saw no way of making all necessary cuts from the domestic budget. Moderate Republicans from the Northeast and Midwest also sought Pentagon reductions, primarily because their regions were hard hit by Reaganomics and not blessed by benefits from military spending.[34] This Republican split was particularly evident in an intraparty squabble over proposed budget freezes. In late 1984 a few conservative Republican House members proposed spending "freezes" that exempted social security and defense, which was allowed 4 percent in real growth.[35] In early 1985, Senator Charles Grassley of Iowa countered with his own freeze proposal that did not exempt defense. He joined with three other Republican members of the MRC, representatives Denny Smith and Tom Tauke and fellow senator Nancy Kassebaum, in producing a paper calling for a military spending freeze.[36] The authors argued that a "reduction in the deficit will contribute more to national security than any other single action." Their efforts reflected widespread and bipartisan sentiments, which were manifested in congressional action on the FY 1986 defense budget. Combined with the impact of the Gramm-Rudman Deficit Reduction Act, Congress actually cut appropriations in real terms for the first time after five consecutive years of substantial growth. The Reagan administration now had more to fear from a bipartisan mixture of conservatives, moderates, and liberals, exemplified by the membership of the MRC, than it did from the peace and arms control movement. Just as the Reagan adminis-

---

flation Windfall," *Congressional Record*, vol. 131, no. 141, 22 October 1985; and Congressional Budget Office, *Budgeting for Defense Inflation*, January 1986.

34. The states of the Northeast and Midwest have the largest negative "balance of payments" with the Pentagon; they pay much more in taxes to the DOD than they get back in contracts, bases, and other DOD spending.

35. Jonathan Rauch, "GOP Conservatives Push Broad Spending Freeze," *National Journal*, 1 December 1984, p. 2283.

36. Charles Grassley et al., *Should We Exempt Defense from Economic Reality?* recommendation to the Senate and House Budget committees and Republican leadership, 7 February 1985. The four authors scored an average 25 percent in their voting ratings by the liberal Americans for Democratic Action.

tration had been compelled to respond to the freeze campaign, it also had to show action on reform.

Reorganizing Waste, Fraud, and Abuse: The Reagan Administration Responds to Reform

The administration could not afford simply to resist reform legislation. The situation required positive action to stem the tide of reform. Whatever the administration had done in the past, the Frank Carlucci initiatives in particular, was politically null and void.[37] Responding to mounting congressional interest in reorganization, Secretary Weinberger fashioned his own reorganization scheme of "management streamlining," which he disclosed in January of 1985.[38] Apparently Weinberger formulated the plan through consultation with only a few associates, including deputy secretary Taft and former deputy secretary Carlucci. One of the principal changes was the creation of an assistant secretary of defense for acquisition and logistics, who would undertake some of the procurement responsibilities formerly with the under secretary for research and engineering. This move paralleled Weinberger's appointment of an assistant secretary for spare parts management when that controversy was raging. Congressional critics charged that creation of this new post had the effect of dividing management of both procurement and readiness, instead of providing greater integration of oversight. Weinberger's plan only added fuel to congressional fire, and some conservatives in Congress urged the president to do something, which he did.[39]

---

37. In April 1981, Carlucci, the deputy secretary of defense, had issued a memorandum to his subalterns in the DOD entitled "Improving the Acquisition Process" and containing thirty-one initiatives, among them measures to increase up-front funding for weapons programs, cut costs, streamline bureaucratic procedures, and make management changes. In July, Carlucci added increased competition. Some accused the initiatives of exacerbating the problems. For example, to get more up-front funding, Carlucci created the larger inflation factor that led directly to the inflation windfall. Reformers also tended to favor centralized management, not decentralization. See Gordon Adams, *Controlling Weapons Costs* (New York, 1983); *Concepts: The Journal of Defense Systems Acquisition Management* devotes the entire Summer 1982 issue to the initiatives.
38. See Michael R. Gordon, "Shifting Boxes," *National Journal*, 7 July 1985, p. 1714.
39. Bill Dickinson, an Alabama Republican and ranking minority member of the House Armed Services committee, was credited with persuading Reagan of the need for action.

In July 1985, Reagan announced he was convening the President's Blue Ribbon Commission on Defense Management, which quickly became known as the Packard Commission because of its chair, David Packard, chairman of Hewlett-Packard, a major defense contractor. Packard had also been deputy secretary of defense under Nixon and in that capacity had overseen an earlier effort to reform procurement during the Vietnam War. Military brass and national security bureaucrats dominated the commission. Four of sixteen members were former Scowcroft commission members, including Brent Scowcroft himself. Independent viewpoints were not represented, least of all by the chair, and conflicts of interest abounded.[40]

Just like the Scowcroft commission, the Packard Commission was an attempt to forge bipartisan agreement through the deliberations of an august assemblage. As with the Scowcroft panel, no one expected any intellectual revolution on the issue of defense management. The purpose of the panel was to show the administration's sincerity on reform and propose a compromise package of reform proposals. The White House hoped to quell congressional concerns and preempt more radical reforms.[41] Members of the MRC worried that the Packard process would snuff the fire they had started.[42]

Feeling the heat from that fire, the President's Blue Ribbon Commission on Defense Management delivered a rather insubstantial twenty-six–page interim report to the president on 28 February 1986. This interim report was followed by more detailed reports in the spring and the final report in June. As noted on the first page of the interim report, the commission had been charged with a substantial task: to study "current defense management and organization in its

---

40. With Lieutenant General (Ret.) Brent Scowcroft representing the air force, two retired generals and a retired admiral covered the army, marines, and navy. Two other notable commissioners were William P. Clark, Reagan's former national security advisor, and Frank Carlucci, the first deputy secretary of defense under Weinberger. The commission staff was also composed of congressional and military bureaucrats.

41. As evidence, John Dingell, chair of the House Energy and Commerce committee, in a letter to the White House quoted a section of NSDD-219, which was the White House order to implement the Packard recommendations. In the quoted section, national security advisor John Poindexter writes that the "Packard Commission got favorable reviews and this gives the President leverage in dealing with reforms on the Hill." According to Poindexter, NSDD-219 "was intended to strengthen your [the president's] hand vis-a-vis legislation on the Hill." Dingell letter excerpted in A. Ernest Fitzgerald, *The Pentagonists* (New York, 1989), p. 259.

42. Winslow Wheeler, a top congressional staffer for leaders of the caucus, said the Packard Commission was the administration's political "master stroke" on reform and likened it to a "steamroller," which would have been difficult if not impossible to stop. Interview with author.

entirety, including: 'the budget process, the procurement system, legislative oversight, and the organizational and operational arrangements, both formal and informal, among the Office of the Secretary of Defense, the Organization of the Joint Chiefs of Staff, the Unified and Specified Command system, the Military Departments, and the Congress.'"[43] The four major areas of study were national security planning and budgeting, military organization and command, acquisition organization and procedures, and government-industry accountability. The commission's task thus encompassed every concern of reformers in Congress. It also was so vast that it perhaps ensured that no single problem would be isolated and adequately addressed.

Despite or because of the vast scope of the project, the commission's findings were predictable, its remedies largely organizational. The report lamented a lack of coherent national security planning and of unified military decision structures. Furthermore, the commission blamed overregulation and bureaucratic red tape for the procurement quagmire: "The nation's defense programs lose far more to inefficient procedures than to fraud and dishonesty. The truly costly problems are those of overcomplicated organization and rigid procedure, not avarice or connivance."[44] Not only was Congress substantially to blame for this, it was also largely responsible for the lack of budgetary and programmatic stability, which the commission also cited as central to the problems of defense acquisition.

The recommendations paralleled the findings: the report urged that two-year budget cycles be used for the Pentagon; that the chair of the Joint Chiefs be given more authority and staff; and that an under secretary of defense for acquisition be created. The headline proposals were kept at the level of bureaucratic organization. Other recommendations for greater use of testing at all stages of weapons development and selectively increased use of competitive contracts were clearly ancillary and unspecified. No mention was made of revolving doors, warranties, or should-cost pricing. In fact, one of the commission's answers to waste, fraud, and abuse was "contractor self-governance," that is, increased reliance on contractors to police their own behavior.[45]

43. The President's Blue Ribbon Commission on Defense Management, *An Interim Report to the President*, 28 February 1986, p. 1.
44. Ibid., p. 15.
45. Ibid., pp. 20–22.

The administration did not wait even for the final report before embracing and endorsing the commission's work and ordering the implementation of its suggestions.[46] The day after he received the interim report, Reagan said he would strengthen the JCS chair, give him independent staff support, and create a vice-chair. He called for the creation of the Pentagon post for oversight of weapons development and procurement. He also recommended two-year budgeting. This order to implement the Packard recommendations became NSDD-219.[47] In a later report to Congress, the president reaffirmed his commitment to these organizational proposals but cautioned Congress not to trample on executive prerogatives and reduce the power of the secretary of defense, referring specifically to the proposed acquisition under secretary as the spearhead of procurement reform: "Beyond this initiative, however, further change to the acquisition organization of the Department of Defense should be left to the Executive branch."[48]

Even so, Congress had been formulating its own version of reorganization while the Packard Commission was drafting its. The leaders of the congressional reorganization efforts were not, however, from the MRC; they were conservative members of the Armed Services committees in both houses. As we have seen, the Armed Services committees had resisted and successfully scuttled or amended several pieces of reform legislation sponsored by members of the MRC, only a few of whom served on those committees. Many members of the Armed Services committees were jealous of their stewardship of defense planning and more conservative in their desire for and approach to military reform. Under the leadership of Barry Goldwater, who inherited the chair of the Senate committee upon John Tower's retirement in 1985, and Sam Nunn, the ranking Democrat, the Senate Armed Services committee made military reorganization a top priority. In the House, Bill Nichols, chair of the Armed Services subcommittee on investigations, started to draft reorganization legislation. The fundamental problems most frequently

---

46. On reasons for Reagan's quick action on the Packard report see Benjamin Schemmer, "What Was Behind Reagan's 'Rush to Judgement' on DOD Reorganization," *Armed Forces Journal International*, May 1986, pp. 15–17.

47. A summary of NSDD-219 is in *A Quest for Excellence: Appendix*, final report of the President's Blue Ribbon Commission on Defense Management, June 1986, pp. 33–37.

48. *Defense Reorganization: Message from the President of the United States*, referred to the Committees on Armed Services, 28 April 1986, p. 7.

discussed at reorganization hearings were various consequences of decentralized authority and interservice rivalry. Reformers perceived an urgent need for greater unity of military advice and command. To back their actions Nunn and Goldwater produced a staff study of reorganization that ran to more than 640 pages. It blamed interservice rivalry for problems in command and planning, and it advocated abolishing the JCS and giving commanders in chief of unified commands greater authority and advisory powers.[49]

Meanwhile the House approved Nichols's bill in November 1985 by a vote of 383–27. The House legislation made the chair of the JCS the principal military advisor to the president and the secretary of defense, gave him an independent staff, and gave greater authority to commanders in chief of the unified commands. The Senate passed similar but more comprehensive legislation based on the Goldwater and Nunn initiative. The Senate bill made the chair the decision maker in the JCS and the principal military advisor to the president and added the chair to the chain of command. It also created a vice-chair who would give field commanders greater input into the planning process. And it called for the creation of an under secretary of defense for acquisition and a cut in DOD staff of 10 percent. The House and Senate continued to vie for the most sweeping legislation through the summer of 1986. The final bill was passed in September as the Goldwater-Nichols Department of Defense Reorganization Act of 1986 (PL 99-433). Despite some differences, the legislation enacted much of the substance of the Packard recommendations.

The Packard Commission was Reagan's solution to attacks on procurement and Pentagon organization. The commission and its work, along with the contributions by the Armed Services committees, effectively conflated both procurement and organizational reform into an overarching problem of management, thus transforming distinct concerns into a package amenable to a narrow set of solutions. The commission's central recommendations all had to do with changing forms of management, and the subsidiary ones were not nearly as specific or urgently recommended. The best example is the procurement czar. Instead of wholeheartedly endorsing competition and should-cost contracting, the panel opted for a managerial solution, which effectively undercut some of the power of the movement's

---

49. The report is titled, *Defense Reorganization: The Need for Change*, staff report to the U.S. Senate Committee on Armed Services, 16 October 1985.

critiques and proposals. In many ways, the triumph of the Packard Commission's approach signaled the beginning of the end of military reform in Congress; which is not to say that the commission's reforms worked or were implemented effectively, if at all.

For example, one of the Packard Commission's principal recommendations was for the establishment of a defense acquisition executive, a so-called procurement czar, in the office of the secretary of defense (OSD). The procurement czar epitomized the commission's attempt to forge managerial solutions to the procurement fiasco by creating a new bureaucratic position. This recommendation was written into the Goldwater-Nichols Act and then amended in the Defense Acquisition Improvement Act of 1986.[50] The acquisition executive was made an under secretary, subordinate only to the defense secretary himself. According to James Woolsey, a Packard member who was on the procurement subcommission, the goal of this office was to effect a fundamental or "cultural change" in the way the Pentagon purchases equipment.[51] The president almost immediately submitted the name of Richard Godwin, a Bechtel executive and long-time associate of Weinberger. Godwin assumed his post on 30 September of that year. Just under one year later, Godwin tendered his resignation. Godwin stated that his "decision to resign rested simply upon my judgement that we were not prepared to move ahead vigorously with implementation of the Packard recommendations and that the institution was not prepared to change the status quo."[52] Given the lofty status of his position within the Pentagon, the acquisition czar was clearly intended to be the focus of decision making with regard to procurement. Furthermore, the presumption was that significant reform of the process was part of the package. Yet Godwin resigned frustrated that "everybody agreed that these changes were desirable so long as we did not make any change."[53]

The House Armed Services hearings into Godwin's resignation did not isolate one particular cause for the failure of the office to produce the desired changes in Pentagon procedures. The study

---

50. Specifically, the act is Title IV of the National Defense Authorization Act for Fiscal Year 1987 (PL 99-661).
51. Woolsey is quoted in U.S. House, Committee on the Armed Services, *Report on the Duties and Authority of the Under Secretary of Defense (Acquisition)*, 100th Cong., 1st sess., 16 November 1987, p. 3.
52. Ibid., p. 1.
53. Ibid., p. 35.

said the causes were "diffuse." After citing Godwin's own inexperience with governmental operations, the study cited under secretary Taft's supervision and its effect of undermining Godwin's authority, Weinberger's failure to go to bat for the new kid on the block, and active resistance by other OSD and service officials, who refused Godwin's requests for data and who were allowed to alter his directives at will. The report emphasized Weinberger's "passive" role, which facilitated bureaucratic resistance.[54]

Weinberger and the Reagan administration, who initially resisted substantive reforms, were perceived, accurately or not, as culpable in the failure of perhaps the most important Packard reform. To derail reform efforts and to muster flagging support for his military buildup, Reagan had bent with the winds of military reform through the Packard process. He managed, however, to limit his flexibility to organizational reforms and even to undermine some of those with potential consequence. Reagan resisted reform legislation until the end of his administration. When Congress passed the Whistleblower Protection Act of 1988, President Reagan killed it with a pocket veto.[55] In a July 1989 report on defense management, President Bush's newly appointed secretary of defense, Richard Cheney, concluded that "efforts to date have not produced the tangible results envisioned by the [Packard] Commission," and that new actions "that substantially depart from or go well beyond DoD's and Congress' efforts to date" are required.[56] Moreover, the effects of other procurement reform legislation, such as CICA and the Defense Procurement Improvement acts of 1985 and 1986, were hard to discern by the end of the Reagan era. Congressional committees apparently had not done oversight reviews of some laws, and other reports drew on data that predated the implementation of the legislation.[57] One investigation done in 1989 by the Senate Governmental Affairs

---

54. Ibid., pp. 3–4.

55. As with CICA, the administration's argument was constitutional. In this case, Attorney General Thornburgh advised the president that the whistle-blower legislation was unconstitutional insofar as it could result in the federal government suing itself.

56. Department of Defense, Dick Cheney, Secretary, *Defense Management: Report to the President*, July 1989, p. 8. A November 1988 GAO report prepared for Senator William Roth found only limited progress in the implementation and impact of the many Packard recommendations.

57. For example, the GAO's *DOD Revolving Door*, GAO/NSIAD-87-116, April 1987, uses data from 1983 and 1984 to show that reporting procedures were inadequate to detect many conflicts of interest.

Committee found that the Pentagon had not increased significantly its use of off-the-shelf or commercial items in spite of the long history of directives and legislation, including an amendment to the 1987 Defense Authorization Act.[58]

By 1987, however, the military reform movement had lost most of its political momentum, although interest in military reform had been growing into 1986. The political capital the issue garnered for such people as Denny Smith and Charles Grassley attracted attention. For example, Grassley's popularity in Iowa increased from about 45 percent to 65 percent in 1983 after press coverage of his reform activities.[59] With congressional elections coming in the fall, military reform seemed to offer a positive and politically viable response to the various problems associated with the Reagan buildup. In the spring of 1986, Smith noted, "We're getting an awful lot of people in the election process that are interested in this. We've had a number [of candidates] contact me and my office about military issues. I think it's not going to go away."[60] Military reform did not go away, but it lost steam after the Packard reforms and never quite recovered. Its influence had ebbed by the 1986 congressional elections. By some measures, the buildup had ended, and other scandals, especially the Iran-Contra affair, soon dominated national attention. The political visibility of reform efforts diminished and legislative reform activity dwindled. Interest in the MRC slackened. When the Levine, Smith, Pryor, and Grassley team had finished its two-year term as caucus chairs, the caucus could not find a Democrat ready to replace Pryor as cochair and Grassley remained on as the Republican. Later the caucus reverted to single chairs from each house. Although military reform became a standard part of the rhetoric on defense policy and played a small role in the 1988 presidential campaign, it never became the basis for a national consensus on defense, as Gary Hart had hoped. Military reform had posed a serious threat both to the Reagan buildup and to the interests of the

---

58. U.S. Senate, Committee on Governmental Affairs, *DOD's Inadequate Use of Off-the-Shelf Items*, report prepared by the Subcommittee on Oversight of Government Management, 101st Cong., 1st sess., 30 October 1989. The hearings and report culminated in the introduction of the Nondevelopmental Items Acquisition Act of 1989 in late 1989. Consideration of the bill began in spring 1990.

59. Charles Mohr, "New Breed of Military Reformer," *New York Times*, 12 October 1983. In fact, Grassley went on to win in 1986 with a substantial 64 percent of the vote, up considerably from his 54 percent in 1980. Jim Andrews's increased popularity back home was also attributable to reform activities.

60. Morrison, "Caucusing for Reform," p. 1602.

arms control lobby and peace groups and thereby added a third side to an already contentious political debate.

Congress and the Politics of Military Reform

Most explanations of the emergence of the military reform crusade in the 1980s see it as a direct response of policy makers to U.S. military performance and international events. These explanations, while credible, are insufficient. Samuel Huntington ascribes reform activities to "the failure of the United States to adapt its military policies to the overall decline in relative American power in the world."[61] Others have cited a related litany of underlying causes, including dismay over thirty years of military defeats, elite dissident defense analysts who publicized institutional problems of the military, the indiscriminate nature of the Reagan buildup, and congress's frustration with its historical inability to influence military policy.[62] What these explanations do not account for is the political context that shaped if not generated the wave of military reform in the 1980s. The movement requires a political explanation. The context had less to do with military failures and new ideas than it did with polarized defense coalitions, domestic politics, and the deficit. The peace movement and the Reagan administration polarized the debate on defense policy. Opinion polls during these years continually showed the public in favor of both a strong defense and arms control. Domestic programs felt the pinch as the guns-versus-butter trade-off became politically important. The deficit climbed to the top of the political agenda. This schism over defense policy and its effects on domestic politics had a profound impact on an already politically fractured Congress.

Congress was not realigned with Reagan and the new conservatism. With the leadership of each house in different partisan hands for most of the Reagan presidency, and with weak majorities in each house, Congress vacillated between New Deal liberalism and the new conservatism. The public perception of the importance of defense issues required senators and representatives who ordinarily

---

61. Samuel P. Huntington, foreword to *The Defense Reform Debate*, Asa A. Clark IV et al., eds. (Baltimore, 1984), p. x.
62. Jeffrey Record, "The Military Reform Caucus," *Washington Quarterly* 6 (Spring 1983): 125–29; James W. Reed, "Congress and the Politics of Defense Reform," in Clark et al., *Defense Reform Debate*, pp. 230–49.

paid little attention to such matters to adopt a coherent position. In the politics of defense reform, many of them found a solution to their dilemma. Military reform converted the struggle between nuclear hawks and doves into one over means rather than ends. Despite moderate tendencies in both camps, for a brief period the clash between the peace movement and the Reagan administration produced a debate about the ends of national security: must the United States engage in perpetual cold war under the nuclear sword of Damocles or does détente and coexistence require arms control, if not disarmament? The sine qua non of the military reform movement was the transformation of the terms of debate from the (nuclear) ends to the (conventional) means of national security. By attacking waste and mismanagement, members could appease the peace movement and make gestures toward fiscal restraint, without being attacked as doves. By not questioning the purpose of the Reagan buildup and not pushing for large reductions in the administration's programs, defense reformers also could claim to be for a strong defense and even take credit for getting more out of every defense dollar. This explains how military reform became a struggle in which Democrats and Republicans in Congress joined together to confront the Reagan administration over defense programs and policies. It also explains the shared skepticism and rancor about the reform movement expressed in the two quotations at the beginning of this chapter, one from a ranking Pentagon official and one from a lobbyist for the peace movement.

Views of Washington insiders notwithstanding, the domestic interpretation of the rise of military reform is difficult to prove because the political consequences of military reform are easier to document than the motivations behind it.[63] It is not easy to distinguish the promotion of a good idea on its merits from political expediency and entrepreneurship—and rarely are the two completely separate. Congressional behavior is often complex, involving a mixture of motivations even within individuals. Probably most reformers came to believe in their cause, but that explains neither its popularity in Congress nor the timing of that popularity. The ideas of military reform would never have generated a national controversy and de-

---

63. This methodological difficulty is shared in two of the most important works on Congress, David Mayhew's *The Electoral Connection* (New Haven, Conn., 1974) and Morris Fiorina's *Congress* (New Haven, Conn., 1977). Both authors, but especially Fiorina, reason from the consequences of congressional behavior back to the motivation they claim must drive the behavior.

bate had not Congress adopted them. The usual explanations account only for the *origins* of ideas, and ideas are rarely politically powerful in and of themselves. What we need to understand is how ideas and the problems that inspire them, such as waste, fraud, and abuse in the Pentagon, are transformed into politically powerful causes and movements. As Les Aspin argued in the 1970s, "Congress is essentially a political institution and responds primarily to political stimuli. Rational arguments in such an institution carry very little weight unless they are politically organized. Political organization can be accomplished, but only with a great deal of effort."[64] As for political stimuli and organization, the controversy surrounding the Reagan buildup provided a powerful dose of the former, while the leaders of the Military Reform Caucus created just enough of the latter. The ideas of military reform would never have generated a national debate had they not found a politically receptive home in Congress.

Military reform, then, served several important purposes for many members of Congress. It provided an organizational principle that brought some coherence to congressional attempts to mediate between conflicting pressures from the public and the executive branch. For individual members, reform ideas and programs became political capital to be invested back in their states and districts in the search for electoral security in a time of political change. These first two benefits were made possible by the keystone of the reform position: the transformation of an ideological or conflictual issue (one characterized by contending and widely held opinions), such as abortion, into an ethical and consensual issue (one characterized by a single position held by the vast majority), such as drugs and crime. Hawks and doves may fight about the efficacy and morality of nuclear deterrence, but who can be for waste, fraud, and mismanagement? With this transformation, the military reform movement changed the tenor and content of national security politics in the 1980s.

---

64. Les Aspin, "The Defense Budget and Foreign Policy," *Daedalus* 104 (Summer 1975): 173.

8

From Buildup to Build-down

By 1987 the nature of the debate and the political competition over defense had changed character. The military budget remained large and a source of controversy and conflict, with advocates of less, greater, and more efficient spending all having their say, but the debate became desultory and sporadic. The buildup, waste, and the danger of nuclear weapons remained, but the passion was greatly diminished. These changes had less to do with the importance of national defense than with the emergence of other issues. Soon after the 1986 elections, the Iran-Contra scandal became front-page news and remained so for the next year. The Iran-Contra affair was the product of the Reagan administration's preoccupation with Nicaragua and the Middle East hostages, two national security concerns for which the military buildup was irrelevant. The White House was in disarray; a second Watergate seemed possible if not probable.

The events of Iran-Contra often overshadowed the apparently earnest arms control negotiations going on between the United States and the USSR. The arms talks were another reason the politics of defense policy were changing. In November 1983 the Soviets suspended negotiations on intermediate nuclear forces (INF), in response to the actual deployment of American Pershing II missiles in West Germany. In December the Soviets also walked out of the Strategic Arms Reduction Talks. It was not until March 1985 that negotiations resumed, this time with Mikhail Gorbachev across the table.

Despite Gorbachev's greater openness and flexibility, and the administration's decreased intransigence, an agreement took more than two years to hammer out. Fitful progress was made in START as well. Reagan's one triumph in foreign policy, the INF treaty, was signed by Gorbachev and the president on 8 December 1987. The treaty (1) forbids the production and flight testing of ground-launched intermediate-range nuclear missiles with ranges between three hundred and thirty-four hundred miles; (2) requires each side to dismantle existing intermediate-range missiles, including all SS-20s on the Soviet side and the American ground-launched cruise missiles and Pershing missiles; and (3) sets precedents for intrusive, on site verification procedures. To the administration, the terms of the treaty vindicated their version of the dual-track approach NATO members agreed to in late 1979: deploy while negotiating. Resisting pressure at home and on European allies, the administration insisted on going ahead with deployment. "Opponents of our deployments have been proven dead wrong," argued Secretary of State George Shultz near the beginning of Senate hearings on the proposed treaty; "This is not the peace movement's treaty. It is NATO's treaty."[1] Speaking of the treaty during the presidential debates that fall, Vice-President Bush said, "Thank God the freeze people were not heard—they were wrong—and the result is we deployed, and the Soviets kept deploying, and then we negotiated from strength, and now we have the first arms control agreement in the nuclear age to ban weapons . . . peace through strength works."[2]

Nevertheless, the peace movement firmly supported the treaty, while many conservatives opposed it. Although it was clear from the start that most Republicans would support their president's treaty, some, including senators Jesse Helms and Dan Quayle, raised concerns and proposed amendments. Helms, Quayle, and their allies expressed reservations about the terms of the treaty and how its implementation would unfavorably alter the overall military balance between NATO and the Warsaw Pact. Some harbored concerns that Reagan was rushing imprudently to complete a START agreement for the Moscow summit in June. They sought to use the INF process to air their concerns about START, if not to delay or alter any possible agreement. Despite such tactics from some of the presi-

1. *Congressional Quarterly Weekly Report*, 30 January 1988, p. 194.
2. "Transcript of the Second Debate between Bush and Dukakis," *New York Times*, 14 October 1988, p. 15.

dent's closest allies, the INF treaty was easily ratified by the Senate in late May, just in time for the Moscow summit. Nevertheless, Reagan's treaty and his new role as international diplomat and peacemaker were somewhat tarnished by the lingering Iran-Contra investigation and its implications about Reagan's "hands-off" presidency.

From Mass Movement to Arms Control Lobby

Even if Reagan could not dispel his Iran-Contra troubles with the INF treaty, the treaty did make a problem for Reagan's once-powerful opposition, the peace movement. The 1984 elections had been one turning point for the peace movement; the INF treaty was another, both substantively and symbolically. The treaty, coupled with Gorbachev's reforms, created a widespread perception, accurate or not, that the corner had been turned in this latest round of cold war belligerence, if not in the cold war itself; for the INF treaty seemed to augur even greater and broader reductions, especially through the anticipated START treaty. The sincerity of Reagan's new internationalism mattered less than the harmonious trio of Gorbachev, *glasnost*, and *perestroika*.

By the time Reagan and Gorbachev shook hands over the INF agreement, the peace movement had experienced its own perestroika, or restructuring. Exhibiting some signs of waning mass commitment before the 1984 election, the peace movement clearly faded thereafter. At its December 1984 meeting in St. Louis, the freeze coalition had in effect acknowledged its declining influence. At this fifth annual conference, the coalition had decided "not to continue as an ad hoc movement, but restructure for the long haul."[3] Having found its appeal too narrow, the campaign had decided to renew its attempt to broaden its mass base, specifically mentioning organized labor and the Rainbow Coalition. But the movement had grown impatient with the results of public awareness and coalition building, so the conference had also made a more important decision to revitalize its efforts at stopping particular weapons programs.[4] Toward

---

3. Marguerite Beck-Rex, news director for the freeze campaign's Washington office, quoted in *New York Times*, 10 December 1984, p. 20; see also *Nuclear Times*, January/February 1985, pp. 7–8 and 11; and "Whither Freeze?" *Nation*, 22 December 1984, p. 669.

4. The delegates, in other words the core activists, seemed to have become more determined in the face of recent setbacks. This led to an interesting mixture of results. Frustration had produced tension between those who wanted to continue to reach

this end in 1985 the organization moved its primary office to Washington from St. Louis, where it had been located at the start to symbolize the grass-roots origins and constituency of the movement. The organization had become less of a mass movement and more of a lobby. Increasing financial difficulties during the next year forced significant decreases in national staff and the consideration of mergers with other peace organizations.[5] By December 1986, agreement had been reached on a merger with SANE under the name SANE/Freeze. In July 1987, William Sloane Coffin became the president of the new organization. But consolidation did not bring renewed strength.

The diminution of the mass movement did not, however, mean the end of influence from the liberal-left on arms control and disarmament. While the mass movement waned in 1984, many of the organizational structures of the peace movement had remained vigorous. Numerous national offices and countless local branches and independent organizations persisted. The Washington lobby was perhaps more powerful than ever and better organized.[6] But the relationship between the well-organized Washington lobby and grass-roots organizations had become somewhat disjointed. And substantive victories remained illusive. Harassment and delays were more common than outright reductions in dollars and weapons. The arms control lobby claimed victory in several setbacks to the Reagan nuclear weapons program. High on its list were capping the MX program at fifty instead of one hundred missiles and compelling adherence to the strict interpretation of the ABM treaty, along with cuts in Star Wars funding and ASAT weapons. Several other victories in the House were subsequently overturned by the Senate.[7] Much of the

---

out to middle America and those looking for more direct tactics. The leadership seemed to move away from its grass-roots origins by moving to Washington and lobbying directly against weapons, a move that seemed to concentrate energy on stopping the arms race. At the same time, the convention endorsed proposals supporting resistance against intervention in Central America and condemning apartheid. It also endorsed participation in certain forms of civil disobedience.

5. Frances B. McCrea and Gerald E. Markle, *Minutes to Midnight* (Newbury Park, Calif., 1989), p. 133.

6. John Isaacs, legislative director for Council for a Livable World, made this assessment in an interview with the author. As David Lewis, lobbyist for Physicians for Social Responsibility, put it, "We've gotten better and better at what we do. . . . We can do more with less"; quoted in Pat Towell, "Bush Era Brings Uncertainty to the Liberal Coalition," *Congressional Quarterly Weekly Report*, 18 February 1989, p. 342.

7. This list is based on a survey of Council for a Livable World mailings from 1985 to 1988.

activity was defensive. Star Wars, which was in part a response to the peace movement, became a target of great effort. The movement had to retool its public relations and legislative campaigns to stop this new threat before it materialized. Defeating the MX had to share time with stopping Star Wars, so SDI and the space weapons debate absorbed some of the resources and energy that could have been devoted to offensive weapons systems.[8]

In this context, the INF treaty was a Pyrrhic victory for the peace movement. Not only did the treaty come well after the movement lost the fight against production and deployment of the Pershing and cruise missiles, but the treaty seemed to vindicate the arm-to-parley approach of the administration. The peace movement and the liberal arms control community took credit for pressuring Reagan into negotiating and tended to applaud Gorbachev's concessions and blame Reagan for delaying the process. But it was Reagan's treaty, accomplished on Reagan's terms, regardless of arguments to the contrary. Reagan would get most of the credit, and the peace movement and arms control interest groups would lose some of their business as public concern eroded. The flow of money from individuals and foundations began to evaporate. The SANE/Freeze merger soon confronted a shortfall in fund-raising. The League of Women Voters dropped arms control from its lobbying activities. Things would only get worse after Bush's election and the Eastern European revolutions in 1989.[9] For example, *Nuclear Times*, the magazine born out of the 1980s freeze movement and chronicler of it, folded involuntarily that year, after foundation funding dwindled. Other peace and arms control organizations experienced similar problems with membership and funding. Some looked for new issues, including nuclear waste and military conversion.[10]

---

8. As noted in Chap. 6, mailings from organizations began to focus on Star Wars. Several new campaigns emerged with fighting Star Wars as their sole object. Even Common Cause, which had jumped on the nuclear bandwagon earlier, made SDI a focal effort. John Isaacs, of the Council for a Livable World, acknowledged that the arms control lobby "had to" commit resources to Star Wars because "it was the most important issue." While agreeing that that commitment came at the expense of attention to other weapons and issues, he argued that the effectiveness of the arms control lobby was "much more expandable than I thought" and could "carry a lot of issues." Interview with author.

9. Said one anonymous former Washington arms control campaigner, "You can't raise money on arms control. It's a new day: You've got an INF treaty and a moderate president"; quoted in Towell, "Bush Era," p. 338.

10. "Those for Whom Peace Dividend Means Deficit," *New York Times*, 12 February 1990, p. 6. Victims included conservative security interest groups as well. *Nuclear Times* managed to find new funding and emerged a year later reorganized as a quar-

Despite a herculean public education effort, the movement accomplished none of its legislative agenda during the first Reagan administration and won only minor skirmishes in its lobbying efforts during the second. This chasm between its apparent potential and its achievements is the paradox of the peace movement. Answers to the failure of the nuclear peace movement will not be found in discussions of its strategy and tactics. The peace movement did not fail because of what it did not do. It failed in spite of all it did do, and because of what it could not do.

What the movement did do was mobilize public concern around the arms race and the threat of nuclear war. This represented a profound departure from traditional public support of cold war policies. Since the Vietnam War a large segment of the American populace has gradually become better informed about, and more skeptical of, the permanent military establishment. The nuclear peace movement was able to translate relatively abstract threats to the public's physical and economic well-being into political action, something the anti-Vietnam movement could not do until the threats were palpable and directed at the white middle class. By derailing the nascent cold war consensus, the movement became the foundation for the liberal counterattack against the president and his policies. While the peace movement could not get its arms control agenda legislated, it did provide the first major focus for opposition to and disruption of the Reagan revolution.

What the peace movement and its allied liberal groups comprising the New Politics coalition could not do was reverse the tide of political change that had been rising since the late 1960s. In an era of Republican realignment and Democratic dealignment, the Democratic party was reluctant to adopt the views of a perceived liberal minority in the face of an increasingly conservative majority. The movement could not institutionalize its power in any governmental (Congress) or quasi-governmental (Democratic party) body. No social movement can achieve much without allies in powerful political institutions. The alliance of the peace movement with the social base and resources of New Politics activism may explain much success the movement had, but it also elucidates the limitations of such efforts in the face of a powerful presidential coalition, a fragmented Congress, and a divided Democratic party.

Finally, the actual legacy of the peace movement on issues of na-

---

terly emphasizing a wider array of issues, including environmentalism and human rights.

tional security was perhaps less tangible than changes in strategic policies or arms control agreements. The peace movement became a vehicle for public education, spreading information and concern about nuclear weapons and the arms race farther and deeper than ever before. Much of that knowledge and concern remained regardless of levels of membership, funding, and activism. In this light, as peace activist Pam Solo put it, "The peace movement does not have the same position in the political debate as it did in the early and mid-eighties. But the peace movement at the local level has never been as strong as it is today. It is a river of opinion, running through the mainstream of society."[11]

Operation Ill Wind and the Irony of Military Reform

If the INF treaty was a partially Pyrrhic victory for the peace movement, the military reform coalition experienced its own dubious triumph near the end of the Reagan presidency. Fading since 1986, reform efforts continued but not with the frequency and power of earlier years. Then, just after Reagan completed the INF treaty and the Moscow summit, a new scandal broke, which would become an ironic legacy of both the Reagan buildup and military reform.

Back in 1986, acting on a tip from an enlisted navy employee, the Naval Investigative Service began an investigation, subsequently joined by the FBI and the Justice Department, into alleged criminal fraud in the procurement process. Attorney General Edwin Meese was not informed. Almost two years later, on 14 June 1988, as part of Operation Ill Wind, agents of the FBI conducted surprise searches, in 12 states, of offices of at least fourteen defense contractors, several defense consultants, and a few Pentagon officials, including high-ranking air force and navy acquisition managers. The contractors included many of the giants of the industry, such as McDonnell-Douglas, Northrop, United Technologies, Unisys, and Litton. Two hundred grand jury subpoenas were issued. The FBI announced that the investigation focused "on allegations of fraud and bribery on the part of defense contractors, consultants, and U.S.

---

11. Marianne Szegedy-Maszak, "Rise and Fall of the Washington Peace Industry," *Bulletin of the Atomic Scientists*, January/February 1989, p. 22.

government employees."[12] In particular, attention focused on what amounted to trading in inside information. The Justice Department alleged that consultants were obtaining classified or confidential information related to procurement contracts from Pentagon employees, through bribes or promises of employment, and then providing that information to a corporate contractor.[13] This inside information could be used to help the contractor make or revise a bid and thus increase the chances of winning the contract. Another accusation was that Pentagon employees had fixed contract specifications to favor a particular contractor. The investigation covered up to seventy-five contracts worth tens of billions of dollars. After the initial coverage following the raids, the investigation quieted down as indictments were prepared.

Like many things, the scandal received scant attention during the election campaign. Whether intentionally or not, federal prosecutors did not bring forward the first indictments until after the elections. Beginning in January 1989, the first indictments were handed down. There commenced a lengthy string of guilty pleas by consultants and corporations. By the end of the year, twenty-six individuals and four corporations had been convicted on charges including bribery and conspiracy to defraud the government. Other corporations, including Boeing and Singer, were hit by investigations unconnected to Ill Wind. In November, Boeing pleaded guilty to two counts of illegally obtaining documents. In fact, one of the first convictions came in a case that preceded Ill Wind. Sundstrand agreed to a guilty plea and a $115 million fine for conspiracy to overcharge the government. Sundstrand officials had padded their contract costs with executive perks including theater tickets, club dues, dog kennel services, and baby sitters.[14] Guilty pleas continued into 1990. In March, Hughes Aircraft, the fifth-largest military contractor, pleaded guilty and was fined $3.67 million for illegally obtaining Pentagon plan-

---

12. Quoted in *New York Times*, 15 June 1988, p. 1. For an overview of the investigation and the types of fraud under scrutiny see David Rosenbaum, "Pentagon Fraud Inquiry: What Is Known to Date," ibid., 7 July 1988, pp. 1 and B5.

13. For example, on 8 December 1989 the Loral Corporation pleaded guilty to three felony charges. The company had paid a private defense consultant $578,000 to persuade two high-ranking Pentagon officials to supply confidential information on contracts for which Loral was competing and to use their influence on the company's behalf. "U.S. Contractor Cites 2 Officials in a Guilty Plea," ibid., 9 December 1989, pp. 1 and 23.

14. *New York Times*, 13 October 1988, pp. 1 and C1; and 14 October 1988, pp. C1 and 4.

ning documents over a seven-year period. By December 1990, twenty-five of the largest one hundred Pentagon contractors had been found guilty of procurement fraud during the previous seven years—some more than once.

The striking irony is that the Justice Department investigation unearthed corruption that had actually been encouraged by mandatory competition in contract bidding. Whereas under noncompetitive contracting, fraud is often associated with overcharging, under a competitive system, contractors may place a high value on getting inside information about other companies' bids and specifications in the scramble to win the contract during the often-protracted competitive negotiations in which bids are adjusted and rewritten. Another gambit is getting the Pentagon to have its specifications favor one company over another. Yet another is collusion among the competing firms. The investigation uncovered evidence of all three. Not only had reformers been generally thwarted in their legislative efforts by conservatives, the Armed Services committees, and the administration, but the little competition they had succeeded in introducing seemed to engender its own form of fraud. In testimony on the scandal before the Senate Armed Services committee, David Packard himself implied that a "misplaced emphasis on competition" fostered some of the fraud the investigation had uncovered. The reforms had contributed to an "environment in which honest and efficient military acquisition is impossible to implement." Paradoxically, Packard also argued that he did not see what harm had been done by one company's alleged purchase of inside information.[15]

The validity of Packard's analysis notwithstanding, one of the strongest reform recommendations in his commission report, contractor self-governance, was looking especially ill conceived and untimely. The Packard Commission had urged that contractors agree to police themselves on procurement ethics and practices. The Pentagon instituted such a program. Forty-six of the one hundred top contractors had enlisted in the voluntary program, agreeing to report violations and make amends without governmental action. Of those forty-six, nearly forty were under investigation either through Operation Ill Wind or other probes.[16]

---

15. U.S. Senate, Committee on Armed Services, *Defense Acquisition Process*, 100th Cong., 2nd sess., 11, 12, 27 July, 4 August 1988, pp. 131–49.

16. Michael Wines, "Conflict-of-Interest Rules Ignored, Pentagon Aides Tell House Panel," *New York Times*, 7 July 1988, p. B5.

Although Operation Ill Wind was the largest investigation of federal procurement fraud in the nation's history, and despite the substantial number of convictions it and other investigations produced, no groundswell for further reform came from Capitol Hill or the White House. Military reform lived on in the determined efforts of a few members of Congress, but its appeal and importance to Congress and the media had diminished. While the objective reasons for reform remained as powerful as ever, the political momentum was gone.

The Reagan Buildup

As both the peace movement and military reform faded from the scene, the Reagan buildup was also changing. The real dollar value of defense appropriations had been falling slowly since 1986.[17] Outlays, or the actual money spent each year, continued to rise through 1987, however, and dropped more slowly than appropriations thereafter. Billions of dollars in appropriations made in long-term contracts during earlier years were coming due as outlays. This "stern wave" of spending would make any decrease in actual spending lag behind cuts in appropriations. Politically, the difference was important because budget deficits are measured in outlays, not appropriations. Regardless, the claim conservatives made that overall defense spending was already declining and that the buildup had ended in 1985 was somewhat fatuous. In the five years from 1981 through 1985, defense budget outlays totaled about 1.346 trillion in 1990 dollars; whereas over the next five years from 1986 through 1990, the total was about $1.553 trillion, amounting to more than a 15 percent increase. Budget reductions notwithstanding, during George Bush's first year in office, defense spending would still be about 40 percent above that in 1980. The buildup continued.

The 40 percent increase over 1980 formed a new and lofty plateau of military spending. Although by 1987 and 1988, many expected larger cuts to be forthcoming, the new plateau had come to be perceived, politically, as a floor rather than a ceiling on defense spending. Few politicians believed increases were likely, but equally few were ready to advocate significant reductions. Anything below current levels still smacked of weakness and neglect. But this political

---

17. By 4.4 percent in FY 1986, 3.0 percent in 1987, 1.9 percent in 1988, and 0.9 percent in 1989, adjusted for inflation. Defense spending as a percentage of GNP was also falling after 1986.

constraint came up against the reality of the deficit and domestic problems that required attention. The Reagan legacy of a 40 percent increase in defense spending and a 20 percent decrease in discretionary domestic programs, coupled with tax cuts and a large deficit, was taking a substantial toll on politics and policy. To many, greater cuts in defense seemed probable, probably inevitable, but not easy.

Defense remained controversial and could be made even more so when politics demanded. In August 1988, Reagan dramatically vetoed the FY 1989 defense authorization bill passed by Congress, something he had never done before. The president objected to several provisions on strategic programs, including cuts in the SDI budget, a delay in the mobile MX program, and other specific arms control restrictions. Reagan argued that the bill tied his hands in arms control negotiations but added a much more sweeping condemnation: "This bill would signal a basic change in the direction of our national defense, a change away from strength and proven success and back toward the weakness and accommodation of the 1970s."[18]

None of these congressional measures was anything new, however. Congress had added many similar stipulations and restrictions to past defense budgets. Large cuts in SDI had been made since 1985. But 1988 was an election year, and many on both sides of the political aisle understood the political motivations behind the veto. Democrats accused the president of deliberately manufacturing the controversy to help Vice-President Bush and the Republicans by once again creating and exploiting perceptions of Democratic infirmity on defense. Republicans charged that the Democrats, because of their politically motivated restrictions, had only themselves to blame for the veto. According to Republican senator Pete Wilson, the veto was "one of those happy times when good government and good politics coincide."[19]

Whether politically inspired or not, Reagan's bit of lame duck bravado seemed a fleeting reminder of days gone by. But even if Reagan and his rhetoric would soon be gone, other peculiarities of the Reagan buildup remained, including Star Wars, which stands apart as the major qualitative military innovation of the Reagan era and, for that matter, of the last two decades. To the end, President Reagan remained committed to his vision of a comprehensive defense.

---

18. For Reagan's remarks and other information on the veto see Pat Towell, "Veto of Defense Bill Ups the Political Ante," *Congressional Quarterly Weekly Report*, 6 August 1988, pp. 2143–45.

19. Ibid., p. 2143.

In 1988, in commemoration of the fifth anniversary of his Star Wars speech, Reagan promised that scientists "are not working late into the night to construct a bargaining chip" and that "we will continue to research SDI, to develop it and test it, and as it becomes ready, we will deploy it."[20] Despite the annual struggle over the program between the administration and its opposition in Congress, Reagan was able to transform SDI into an ongoing three or four billion-dollar-per-year program. In October 1988, less than a month before the elections, the Defense Acquisition Board approved plans for development of a Phase I Strategic Defense System. This partial defense would be composed of kinetic weapons based in satellites and terminal defenses on the ground. No lasers would be used. The projected cost, according to the Pentagon, would be seventy-four billion dollars. In the days and years to follow, however, someone else would have to deal with Star Wars as well as the broader consequences of Reaganism and the military buildup.

The Elections of 1988

In spite of the tone set by Reagan's veto, issues of national security and defense policy did not figure prominently in the presidential campaign of 1988. Then again, few serious issues did. Both parties offered far more restrained rhetoric on the subject of national security and defense. Both candidates talked in generalities about the need for a strong but efficient defense.

The Democrats and Dukakis tried once again to steer their party down the middle. Part of the strategy was embodied in the very short and epigrammatic party platform. Long on principles and short on programmatic proposals, it reflected Democratic anxieties about appearing to be anything in particular. The document was especially vague about national defense. Its opening statement on defense policy argued, "Our national strength has been sapped by a defense establishment wasting money on duplicative and dubious new weapons instead of investing more in readiness and mobility." Instead, security would "be enhanced by more stable defense budgets." Whether stability meant cutting, adding, or zero growth was not specified.[21]

On the campaign trail, Dukakis and Bentsen tried to be tough on

---

20. Lou Cannon and R. Jeffrey Smith, "President Blames SDI Delays on Hill's 'Irresponsible' Cuts," *Washington Post*, 15 March 1988, p. 6.

21. The 1988 Democratic platform is reprinted in *Congressional Quarterly Weekly Report*, 16 July 1988, pp. 1967–70.

issues of defense and national security. The ticket pledged to maintain a stable defense budget, which meant, to Bentsen at least, "no cuts in defense—none." On the one hand, Dukakis tended to ridicule Star Wars and criticize the MX and Midgetman programs. On the other hand, he supported the Trident D-5 missile, the advanced cruise missile, and the B-2 Stealth bomber, saying at one point that the "bomber was started by a Democratic President, it was supported by a Democratic Congress and it's going to be completed by my Administration."[22] The governor's defense platform was not centered on the Stealth bomber, however. His emphasis clearly was on conventional forces: "We have a strong and survivable nuclear deterrent. . . . But we continue to have serious deficiencies in our conventional forces."[23] Dukakis called for a "conventional defense initiative," which would include "applying advanced technology to the challenge of fighting and winning a conventional war."[24] He also threw in tough talk about Pentagon fraud and waste and tax dollars "lining the pockets of dishonest defense contractors and consultants."[25]

Besides statements about staying the course and maintaining the buildup, Bush tried to match Dukakis's push for stronger conventional forces with the idea of "competitive strategies," which amounted to emphasizing America's comparative technological advantages rather than matching the Soviets tank for tank. Although Bush characterized this as a new idea, it was little more than a reiteration of long-standing American military policy.[26] The vice-president also gave a strong endorsement to the then two-year-old Packard Commission recommendations, some of which had been implemented and proved lacking, such as the procurement czar. Bush believed the Packard recommendations could save money in the defense budget.[27] Savings aside, Republican commitment to in-

22. "Dukakis Stresses Defense and Ridicules Bush and Quayle," *New York Times*, 13 September 1988, p. 10; see also "Transcript of the First TV Debate between Bush and Dukakis," ibid., 26 September 1988, p. 18.
23. "Dukakis Seeks to Assure Allies on Support for NATO," ibid., 15 June 1988, p. B6.
24. Ibid.
25. "Dukakis Stresses Defense."
26. During the presidential debate, Bush characterized "competitive strategies" as a "very sophisticated concept" and said that "it is new and it is very, very different than what's happened. But it's not quite ready to be totally implemented." From "Transcript of the Second Debate between Bush and Dukakis," *New York Times*, 14 October 1988, p. 15.
27. Ibid.

creasing defense budgets was forthright but tempered: "Continued economic growth will allow more dollars to be available for defense without consuming a larger portion of the GNP or the federal budget; continued efficiency and economy will assure those dollars are well-spent."[28] Bush and his party also remained committed to SDI. The party platform stated that the Republican party was "committed to rapid and certain deployment of SDI . . . ," which is the "most significant investment we can make in our nation's future security."[29] Bush proclaimed, during the presidential debate, "I will research it fully, go forward as fast as we can. . . . And when it's deployable, I will deploy it."[30]

Even though defense was not a focal issue, Bush's comparative advantage over Dukakis was greater with defense than the economy, according to some poll results. In one September survey, Bush was virtually tied (46–44 percent) with Dukakis among potential voters on which candidate was better able to handle the economy. When it came to which candidate was better able to maintain a strong defense, 60 percent chose Bush versus 27 percent for Dukakis.[31] On election day, George Bush won yet another decisive victory for the Republicans. By many measures, Bush won on the basis of his experience. One election day poll found that 46 percent of Republicans made their choice on the basis of candidate experience. Despite the lack of substantive discussion of most issues, defense was important and decisively favored Bush. Twenty-seven percent of Republicans listed national defense as the issue on which they most liked the stand of their candidate. The closest competitors were the national economy (19 percent) and taxes (15 percent). National defense mattered most to only 7 percent of Democratic voters.[32]

Meanwhile the Democrats, once again, licked their wounds and drew familiar lessons. Interviews by Christopher Matthews with several Democratic leaders after the election showed that these leaders considered national defense the issue on which Democrats continued to founder. They described defense as an "overriding" or

---

28. The 1988 Republican platform is reprinted in *Congressional Quarterly Weekly Report*, 20 August 1988, pp. 2369–99.
29. Ibid., p. 2395.
30. "Transcript of the First TV Debate," p. 18.
31. NBC News/*Wall Street Journal* poll data, cited in "Bush Leading Dukakis in Two Polls: Defense Linked to G.O.P Gains," *New York Times*, 23 September 1988, p. 12.
32. All data taken from ABC News/*Washington Post* election day survey, reported in *Public Opinion*, January/February 1988, p. 33.

"threshold" issue that has obstructed if not prevented Democratic victory. Even a liberal former congressman like Mike Barnes of Maryland concluded, "We should have ran [sic] the 1988 campaign the way we did in 1960. We needed to do the same thing that Jack Kennedy did with the missile gap: come at the Republicans from the right."[33] Once again, much of the Democratic leadership concluded that the political breezes should push their party to starboard.

But neither the public nor most members of Congress were willing to sustain the military buildup. The Democrats had increased their majorities in both houses. The split-level realignment of divided government continued. Congress and the president would once again be locked in conflict. Many academics, commentators, and newspaper editors sought a rational explanation and found one in a principle as old as the republic. The people, some said, had made an addition to the inventory of checks and balances. Voters were electing Republican presidents to run foreign policy and control spending and Democratic representatives to attend to local concerns and bring home the federal bacon. Whatever the merit of this behaviorialist explanation, the all too real problem of governmental deadlock remained. Bush knew there would be revisions in the Reagan Great Equation, and defense spending would be near the top of the list.

Old Business and New Breezes

Even before the revolutionary changes in Eastern Europe, the Reagan fiscal legacy, especially the persistent problem of the deficit, compelled policy makers to trim bits and pieces from the defense budget increases still being planned at the end of the Reagan administration and as Bush took office. These initial attempts to adjust military spending to budgetary realities provided indications of how difficult retrenchment might be.

Finally implementing in 1988 a suggestion made in 1984 by the President's Private Sector Survey on Cost Control, better known as the Grace Commission, Defense Secretary Carlucci formed the Defense Secretary's Commission on Base Realignment and Closure. Military bases had been an acknowledged source of Pentagon inefficiency for many years. Knowledge did not mean action, however.

---

33. Christopher Matthews, "Democrats Watch Inaugural from Afar," *San Francisco Chronicle*, 22 January 1989.

No base had been closed since 1977 and for distinctly political reasons. Whatever list the defense secretary would submit to Congress was torn asunder by pork-barreling and logrolling, which so often protect the district benefits of individual members. To surmount this obstacle, the Grace Commission had recommended a presidential commission and revised approval procedures. Carlucci's commission was, as always, composed of a bipartisan combination of politicians, retired military officers, and businessmen. When in late December 1988 it submitted its list of eighty-six bases to be closed and fifty-four to be realigned, Carlucci had to accept or reject the list as given within fifteen days. If he approved the list, Congress had forty-five days in which to disapprove the list as a whole. This procedure was intended to circumvent congressional politics and force a positive decision. The procedure worked, but not without some difficulty. Congress took all forty-five days, during which senators and representatives, liberals as well as conservatives, whose states and districts would suffer, made every effort to stop the recommendation or alter the outcome. For example, Senator Alan Dixon, whose state, Illinois, would lose two large bases, tried to get access to the classified transcripts from the commission's deliberations. Liberal Democrats Nancy Pelosi and Barbara Boxer argued that closing the Presidio, San Francisco's parklike home to the Sixth Army, would cost more than keeping it open.[34] Opponents of the plan threatened to keep their bases open through the appropriations process; for Congress would have to appropriate an estimated five hundred million dollars in each of the next two years to fund the closures and realignments. Members could attempt to keep their bases open by attaching amendments to the appropriation stipulating that no money could be spent to close this or that installation. Although both the House and Senate voted by large margins to accept the base-closing plan, efforts to change some of the specifics continued. This episode revealed the kind of politics and obstacles that were to confront any attempts to decrease defense spending.

Bush came into office touting his solution to the budget crisis, the "flexible freeze." Under this arrangement, most federal programs would be frozen at the previous year's dollar levels. But the freeze

---

34. By law, the Presidio could not revert to private property; instead, it was to be absorbed into the Golden Gate National Recreation Area. According to Pelosi and Boxer, conversion and management under the Interior Department would in fact cost more. Neither representative was willing or able to defend the military value of the Presidio.

did not apply to defense; it would be allowed to keep pace with inflation in FY 1990 and then rise by 1 percent per year beyond inflation in 1991 and 1992, then 2 percent thereafter. Bush's plan was quickly dubbed the "flexible squeeze" for what it would do to domestic programs favored by Democrats. To make the Gramm-Rudman deficit target for FY 1989, the bulk of the cuts, amounting to nearly 10 percent off of $136 billion, would have come from discretionary domestic spending, which was just 12 percent of the entire budget. Democrats in Congress caught in the squeeze would have to make the choices or go against Bush by raising taxes or making large cuts in defense. Although the flexible freeze would disappear from the president's lips and White House planning, the general fiscal squeeze remained the central dilemma confronting the new government, in part because there were problems that could not be frozen or squeezed away.

Over the ensuing early weeks of 1989, the hearings on Bush's nominees for Cabinet posts served as reminders of these constraints. Richard Darman, who was to manage the Office of Management and Budget (OMB), could be heard bemoaning the deficit and uttering the refrain that was on the lips or in the mind of many politicians: No New Taxes. Darman's tough talk on taxes contrasted with the litany of domestic problems either deferred or aggravated during the Reagan era being discussed in other hearing rooms. Before one committee, Jack Kemp could be found expounding on the degeneration of the department of Housing and Urban Development and the sorry state of low-cost housing and homelessness throughout the country. Seated in front of another, Admiral James Watkins, the nominee for the Department of Energy, talked about the crisis in the nuclear weapons production industry. All the major tritium-producing facilities had been shut down, and other nuclear materials facilities were under investigation for significant problems. Cleanup and new construction would cost billions—anywhere from fifty to two hundred billion dollars over twenty years. Watkins also planned to reduce the nation's dependence on oil (which had grown significantly under Reagan) while fighting global warming. William Reilly, the nominee to be administrator of the Environmental Protection Agency, was quizzed about problems ranging from toxic waste to global warming and other emerging environmental dilemmas. The Department of Education hearings were a reminder of the despair in lower education and expense of higher education. Even though conservatives concentrated on the nominee's personal position on abortion, the AIDS crisis and the broader problem of national health care costs loomed large at hearings on the nomination of Louis W. Sullivan to be secretary of health and human services. The savings and loans crisis was continuing to grow; early, and optimistic, OMB estimates of the bailout costs came to $148 billion over ten years, with taxpayers contributing $60 billion. Finally, before his personal habits and conduct displaced national security as the focus of his confirmation, defense secretary designate John Tower, formerly one of the most staunch defenders of the military buildup, acknowledged that the Pentagon would have to live within constraints and might have "to try to achieve as much if not more defense for less money."[35]

Indeed, Tower's more successful replacement, Defense Secretary Dick Cheney, emerged in late April with a revised FY 1990 military budget that was ten billion dollars smaller than Reagan and Carlucci's original version submitted that January. These reductions were achieved through cutbacks and slowdowns in numerous programs, including the retirement of ships and cancellation of plans for a fifteenth aircraft carrier. Cheney even terminated a couple of programs outright. The Democratic majority in Congress had long wanted this kind of realism, and it must have seemed a welcome relief from the intransigence of the Weinberger years. To some, however, it was the wrong relief. In particular, Cheney's decision to end further production of the Grumman F-14D naval air superiority fighter and cancel the Osprey V-22 aircraft program provoked vigorous counterattacks by bipartisan coalitions in Congress. Long Islanders, including liberal Tom Downey and conservative Republican Norman Lent, teamed up to save the F-14, or more accurately, the Grumman Corporation. This group went so far as to use a video of the popular movie *Top Gun* to help persuade fellow members of the plane's vital importance to national security.[36] The tenacious efforts paid off in a fight that continued through the summer and well into the fall. Negotiations between congressional conferees and Cheney resulted in what those involved called a "soft landing" for Grumman: production would continue for eighteen more aircraft and then the program would end.[37] In perhaps an unintentionally revealing

35. U.S. Senate, Committee on Armed Services, *Nomination of John G. Tower to Be Secretary of Defense*, 101st Cong., 1st sess., 25, 26, and 31 January, 23 February 1989, p. 30.
36. On the F-14 lobby's efforts see *Congressional Quarterly Weekly Report*, 27 May 1989, pp. 1268–70.
37. Senator John Warner remarked that he had "never seen such forceful, if not

commentary on the F-14 battle, Lent characterized House support of further production as "a great victory for Long Island and the Grumman employees, and I think for national security as well."[38] Development funds for the Osprey were also reinstated. Unlike pork barrel and soft landings, hard decisions were in short supply.

The politics of military policy could get even more Byzantine when difficult choices on more expensive programs were at issue. For example, the MX and Midgetman missiles remained a perennial problem from the Scowcroft compromise onward. The compromise had created no consensus; land-based missile programs remained as controversial as ever. Deployment of the MX in fixed silos had been capped at fifty by Congress; the single-warhead Midgetman, or small ICBM, was still under development. Although the air force had always been against Midgetman, a coalition in Congress, led, ironically, by moderate and liberal Democrats, had kept Midgetman alive. Cheney recommended, as part of his cuts, putting the fifty MX on railcars and terminating the Midgetman program. Although Bush, whose national security advisor was Brent Scowcroft, would soon overrule his defense secretary, Cheney's plan provoked shock and opposition in Congress. Both Sam Nunn and Les Aspin were staunch adherents of the Scowcroft compromise and the Midgetman missile.[39] Odd alliances were formed. Some liberals wanted to cancel MX and support Midgetman. The MX would cost several billion dollars over the next few years, while the annual development budget of Midgetman was still a mere one hundred million dollars and would increase only gradually for several years. This option would help with the deficit in the near term while leaving open the possibility that Midgetman could be obviated by arms control agreements before deployment. Other liberals such as Barney Frank joined conservatives who supported the largely completed MX program and thought that Midgetman would drain money away from future needs. Of course other conservatives were happy to spend money on both missiles. The result was all too predictable. Bush decided to go ahead with both systems and Congress complied. With Demo-

---

ruthless, lobbying." For this and other details on the F-14 and Osprey compromises see ibid., 4 November 1989, p. 2964.

38. Quoted in *New York Times*, 28 July 1989, p. 9.

39. As Nunn put it, "MX in the last two or three years has been riding on the back of Midgetman." *Congressional Quarterly Weekly Report*, 22 April 1989, p. 916.

crats and Republicans in effect cooperating to perpetuate both ICBM programs, room for reductions was hard to find.[40]

Given the moderate pace of Soviet reform up to that point, it looked like defense would continue to decrease in importance but gradually. As Bush, Cheney, and Congress squared off over the details of a marginally leaner defense budget during the spring and early summer of 1989, events in Eastern Europe only hinted at the revolution to follow that fall. The central theme of Bush's inaugural address in January had been the "new breeze" of freedom blowing around the world. Little could Bush or the Democrats know just how rapidly the breeze they felt would become a political hurricane sweeping aside many assumptions of cold war politics. Beginning with liberalization in Poland and Hungary, the revolutions in Eastern Europe accelerated as the year progressed. The most significant developments came in October and early November, when the East German government fell and the Berlin Wall was opened, thus knocking out the keystone of the Eastern bloc. The revolutionary process continued right up through Christmas, with the deposition and execution of Romania's Nicolae Ceausescu and the selection of Vaclav Havel as Czechoslovakia's president on 29 December.

The fall of the Berlin Wall was followed by a similar tumble in projected defense budgets, although the relationship between the two events was not so direct. In mid-November, Cheney ordered the military services to formulate plans for cutting $180 billion from previously planned increases in the defense budget from FY 1992 through 1994. The potential reduction was not as large as it sounded because it was coming off the top of the original five-year plan that projected steady growth over and above inflation.[41] Nevertheless, if fully implemented, the cuts would have amounted to about a 5 percent per year real reduction in defense spending, depending on inflation. Although the planning exercise was termed a "worst-case scenario" by one Pentagon official, Cheney insisted, "If we do not

---

40. Michael R. Gordon, "Officials Say Bush Will Back a Force of Mobile Missiles," *New York Times*, 22 April 1989, p. 1; Susan F. Rasky, "Coalition Opposes Deal on Missiles," ibid., 24 May 1989, p. 18.

41. The misperception of large cuts provoked several attempts to show how small the real reduction would be, including one by the chair of the Senate Budget committee, Jim Sasser: "The Peace Dividend: A Rubber Check," ibid., 19 December 1990, p. 21.

take the bull by the horns we will be gored by it."[42] Bush had originally planned a 1991 military budget of $311 billion, a slight increase after inflation over the 1990 figure. By late December the figure had dropped to about $307 billion, a small but real decrease after inflation. Even many conservatives thought that would not be enough. Senator James Exon said, "There is a general perception in the land that peace has broken out all over. The public will demand, and the Congress will respond in some fashion."[43] In less than one year the Bush administration had gone from planned increases in the military budget of 1–2 percent to potential 5 percent decreases.

How much of this had been driven by international events, as opposed to the fiscal legacy of the Reagan Great Equation was an open question. Sam Nunn's assessment was that "Secretary Cheney's orders to the services reflect fiscal change, not a real threat assessment."[44] According to Nunn, while the threat had decreased, the emerging changes in Pentagon planning were more a product of budgetary realities. In fact, Budget Director Richard Darman had been fighting Cheney for further cuts in the military budget before either the fall of the Berlin Wall or the defense secretary's call for reductions. The consequence of the domestically motivated search for large cuts in advance of any revision of military strategy was that, in the words of one administration official, "Defense policy is in total chaos."[45]

While the Pentagon raced to make strategic planning catch up with budgetary constraints and international events, plans for significant and long-term reductions gained credibility never imagined in the cold war era. Cheney's worst-case planning was no match for what some analysts had in mind. For example, William W. Kaufmann, a veteran defense intellectual and long-time advocate of a leaner defense establishment, issued a plan for a nearly 50 percent cut in the military budget, to be implemented over the next ten years. Kaufmann's plan, among others, received serious attention and favorable press.[46]

---

42. Michael R. Gordon, "Military Ordered to Draft a 5% Cut in 1992–1994 Spending," ibid., 20 November 1989, pp. 1, 8.
43. Ibid., p. 8.
44. Susan F. Rasky, "Lawmakers Criticize Cheney on Cuts," *New York Times*, 27 November 1989, p. 9.
45. Anonymous, quoted in John D. Morrocco, "Defense Cuts May Force Trade-off between New Systems and Upgrades," *Aviation Week and Space Technology*, 4 December 1989, p. 22.
46. For examples of favorable coverage by mainstream press see Leonard Silk, "A

As early as February 1990, the three most recent former chairs of the Joint Chiefs of Staff, all three of whom had served during the Reagan buildup, appeared before the Senate Armed Services committee in general agreement that most proposed strategic nuclear weapons programs should be reassessed if not terminated. John Vessey and William Crowe argued that national defense did not require two new land-based missile systems; in fact, one could be canceled and the other traded away in arms control negotiations.[47] In March, William Webster, director of the CIA, testified before the House Armed Services committee that changes in Eastern Europe and the Soviet Union "were probably already irreversible in several critical respects." Liberals embraced Webster's measured but definite diminution of the official threat. Cheney responded that Soviet military power remained undiminished and political change could be reversed quickly.[48] In fact, the administration remained committed to much of the buildup, including its two most expensive and, to many critics, increasingly irrelevant items, the B-2 bomber and SDI. George Bush had inherited an SDI that was deeply rooted, if not quite flourishing. In the first budget he prepared as president, he requested, unsuccessfully, a 21 percent real increase in funding for Star Wars. Over the next five years the administration planned to spend an additional thirty-one billion dollars for SDI research and development. In spite of congressional cuts and global change, the administration defended the proposed increases for FY 1991 and beyond. Even Budget Director Darman justified the funding by saying, "There are third parties now, non-superpowers, who have ballistic missile capability."[49] Soon thereafter, President Bush gave a speech

---

Master Plan for Military Cuts," *New York Times*, 24 November 1989, p. C2; Paul Mann, "Brookings Study Suggests Halving U.S. Defense Budget in Ten Years," *Aviation Week and Space Technology*, 27 November 1989, pp. 19–20. Kaufmann's plan was later published as *Glasnost, Perestroika, and U.S. Defense Spending* (Washington, D.C., 1990). *Time Magazine*, mentioned Kaufmann's plan and provided its own blueprint for cutting billions off of Bush's proposed cuts; see "How Much Is Too Much?" 12 February 1990, pp. 16–21.

47. Steven A. Holmes, "Mobile-Missile Curb Favored by Three Ex-chairmen of Joint Chiefs," *New York Times*, 3 February 1990, p. 7.

48. For the Webster-Cheney debate and an example of liberals' embrace of Webster's assessment see Tom Wicker, "Cheney vs. Webster," ibid., 24 March 1990, p. 25.

49. "Budget Director Defends Military Spending Plan," ibid., 29 January 1990, p. 10.

in which he argued, "Strategic defense makes more sense than ever before."[50]

But Bush, Cheney, and some congressional conservatives were fighting an uphill battle against the perception that, as Senator Exon put it, "Peace has broken out all over." Public opinion had in fact undergone a dramatic reversal. Support for the buildup rapidly eroded while support for domestic programs steadily increased. Well before the 1984 elections a majority of the public favored cutting defense spending to lower the deficit, and a consistent plurality believed the United States was spending too much on the military. Both percentages increased slightly during the second Reagan administration.[51] Conversely, support for increased social spending grew to a plurality and in some areas to a substantial majority. By the time George Bush became president, public opinion was the inverse of what it had been in the late 1970s and the very early 1980s, when pluralities favored raising defense spending and cutting most domestic programs. In polls in 1989 and 1990, defense ranked dead last on the public's list of governmental spending priorities (Table 12). In May 1990, when asked what budget cuts were acceptable, 64 percent considered defense reductions acceptable and 30 percent thought them unacceptable; wheras 86 percent viewed cuts in education as unacceptable, and 74 percent thought the same about environmental programs.[52]

The United States was winning the cold war but losing the budgetary war. The obvious conclusion was that deep cuts in the defense budget were inevitable. Ironically echoing Bush's inaugural metaphor, Republican senator Ted Stevens remarked, "This is not just a wind blowing, but a gale coming at the defense budget."[53] Bush cautioned that it was too early to cut defense because the rapid course of change in the Eastern bloc remained unpredictable. Invoking a new metaphor, he likened the changes to a bridge from cold war to lasting peace, which he did not want to cross before being certain "the bridge is secure."[54] Nevertheless, the apparent inev-

---

50. "President Urges Caution in Reacting to East Bloc," *San Jose Mercury News*, 8 February 1990, p. 1.

51. For longitudinal data see *Gallup Report* 260 (May 1987): p. 3; and ibid. 263 (August 1987): p. 24.

52. CBS News/*New York Times* poll results reported in *New York Times*, 27 May 1990, p. 13.

53. Rasky, "Lawmakers Criticize Cheney on Cuts."

54. Remarks on a West Coast trip, during which he visited war game exercises and

Table 12. Public priorities for government spending, October 1989 (in percentages)

	Increase	Same	Decrease
Public education	76	19	3
Help homeless	71	20	6
Reduce air pollution	59	30	9
Assist low-income families	56	31	10
College financial aid	44	39	15
Child-care services	41	42	13
Space exploration	21	39	38
Defense spending	14	41	42

Source: *Gallup Report* 289 (October 1989): p. 6.

itability of large cuts prompted a debate over what came to be known as the "peace dividend," the anticipated savings from defense cutbacks. Just as most liberal and conservative politicians in the late 1970s and early 1980s did not fight over whether the defense budget would grow but by how much it would grow, by the beginning of 1990, many of these same politicians did not fight over whether the defense budget would be cut but by how much and, perhaps more important, what would be done with the savings. While President Bush and Budget Director Darman cautioned that a peace dividend did not yet exist and would be smaller than assumed, many liberals, including senators Cranston and Kennedy, called for reinvestment of the dividend in neglected domestic programs and problems.[55] For some, the war on drugs was a popular substitute for the cold war. But many conservatives and liberals alike agreed that the first priority should or would be the deficit. As Les AuCoin said of the deficit and the dividend, "We've got a fire-breathing monster staring at us. . . . There's just not going to be much of a dividend here to be spent elsewhere."[56] Other conservatives thought tax relief was in order; Senator Phil Gramm's argu-

---

inspected a B-1 bomber and Star Wars laboratories; *New York Times*, 9 February 1990, p. 7.

55. Gerald F. Seib, "Conservatives Argue over Best Ways to Spend a Peace Dividend, and Whether There Is One," *Wall Street Journal*, 31 January 1990, p. 18; David E. Rosenbaum, "Two Senators Say Peace Bonus Should Pay for Social Needs," *New York Times*, 14 April 1990, p. 8.

56. David E. Rosenbaum, "Pentagon Savings Unlikely to Be Spent Elsewhere," *New York Times*, 22 November 1989, p. 14. Another Democratic advocate of using the dividend for deficit reduction was Senate Budget committee chair Jim Sasser.

ment—"I don't think the benefits from winning the cold war should go to the Government"—was hardly unique.[57] Much of the public, however, did not seem to share Gramm's opinion. A majority (61 percent) favored using the peace dividend to fight domestic problems rather than cut taxes (9 percent) or cut the deficit (23 percent).[58]

As 1990 began and the debate over the FY 1991 defense budget ensued, Congressional leaders sought further reductions and the military services made plans for smaller forces.[59] Secretary Cheney began bargaining over specific weapons program reductions by proposing substantial cuts in three major aircraft programs, including a 43 percent reduction in Stealth bomber procurement.[60] In a move that produced consternation among many Democrats, Cheney also proposed more base closings. As the year progressed, the Bush administration and the Pentagon consolidated plans for a substantial retrenchment or "build-down" in defense, during which significant weapons modernization would proceed while the overall size of military forces would be reduced about 25 percent by the middle of the decade. Included in the planned cuts were ten divisions from the Army, ninety-four ships from the Navy, and ten tactical wings from the Air Force. The largest reductions would be in the number of personnel, both military and civilian. Spending would decrease accordingly, falling by 1996 nearly to the purchasing power of the 1980 budget.[61] One complication was that the restructuring proposal, detailed in Bush's FY 1992 budget, was written during the deployment of American forces following the August invasion of Kuwait by Iraq. Economic recession, an ongoing budget crisis, increasing Soviet instability, and a war in the Persian Gulf made such long-term planning and budgeting rather suspect. Despite the apparent inev-

57. Ibid. Or as House minority whip Newt Gingrich argued, "It's not a peace dividend. . . . It's a peace non-expense"; quoted in Susan F. Rasky, "A New World with New Issues Born in the Recess is Awaiting Congress," *New York Times*, 22 January 1990, p. B7.
58. CBS News/*New York Times* 30 March–2 April poll, obtained from CBS News.
59. The president proposed a military budget of $306.9 billion, including military programs in the Energy Department, a 2 percent cut adjusted for inflation. The budget passed by Congress in October cut $18.6 billion from that for a final defense authorization of $288.3 billion.
60. Cheney proposed a cut from a total of 132 B-2s to 75. The plan drew immediate criticism because the 43 percent cut in the number of planes would save only about 20 percent of the projected cost of the original program.
61. For an analysis of the projected force reductions and fiscal implications of the plan, see Congressional Budget Office, *An Analysis of the President's Budgetary Proposals for Fiscal Year 1992*, March 1991, pp. 57–83.

itability of reductions, the exact future of the defense budget was uncertain, and a peace dividend remained an elusive objective.

The Political Legacy of the Reagan Buildup

Whether the cold war was really ending remained to be seen. Nevertheless, Senator Gramm's remarks about "winning the cold war" reflected a common assumption, however premature. All but the most cautious believed the United States had won the decisive round if not the whole bout. But how was this victory, whatever its scope and durability, achieved? There was surprisingly little serious reflection on the relationship between U.S. policy and change in the Soviet Union and Eastern Europe. The most commonly expressed answer among elite opinion makers was some version of peace through strength, with credit often given directly to the Reagan buildup.[62] Fewer were willing to echo Strobe Talbott's assessment that the rapidly proceeding self-destruction of communism demonstrated that "the doves in the Great Debate of the past 10 years were right all along."[63] The changes were undeniable, but analysis of what caused them and interpretation of their meaning seemed to be shaped by ideological and political predispositions. Instead of any sustained reflection on how we got here, the only question that seemed to matter was, Where do we go from here?

What role *had* the defense buildup played, if any, in bringing about international change? As people argued about the existence, size, and purpose of a peace dividend, few bothered to question the costs and consequences of what was, by some measures, the largest and most sustained postwar military spending increase. In 1978, a pivotal year politically in the "decade of neglect," budget outlays for national defense were $104.5 billion or $212.5 billion when adjusted for inflation (FY 1991 dollars). In 1987, the apex of the Reagan

---

62. In defending his 1991 budget, Cheney argued that congressional cuts could undermine the defense strategy that "has directly contributed to the collapse we see in the Soviet Union." *San Jose Mercury News*, 30 January 1990, p. 6. Jeane Kirkpatrick offered a more nuanced assessment of the impact of Reagan and the buildup: "One was the confident reaffirmation of the case for free institutions. And two I think that the restoration of military strength, occurring as it did with the stagnation of the Soviet economy, really dramatized and illustrated for Gorbachev the problems in that country of technology and economic base"; quoted in *New York Times*, 14 January 1990, p. D24.

63. Quoted in Maureen Dowd, "A Sense of Wonder among the Mere Bystanders," *New York Times*, 8 February 1990, p. 9.

buildup in terms of outlays, the military consumed $282 billion ($325 billion in FY 1991 dollars). The cumulative difference between the 1978 figure and all eight years of the Reagan presidency is $1,043 billion, or $636 billion in inflation-adjusted dollars. Add 1980, when Congress and Carter made a real increase, and 1989, and we arrive at a decade's total of about $1,271 billion or over $760 billion when adjusted for inflation.

The $760 billion question is, Did this expenditure make any difference in world politics? Did the nation receive its money's worth, either in increased military might or in enhanced ability to prevent or mitigate threats to national security? If nothing else, the military reform movement showed that the buildup was wasteful and inefficient, often fraudulently so. By many measures, there was a yawning disparity between how much was paid and what was obtained. Another consideration is how irrelevant the buildup was to many of the country's most important security concerns of the decade, including Afghanistan, Lebanon, the Persian Gulf, and Central America.

This leaves the possible link between the buildup and political change in the Soviet Union to examine. The problem is, of course, that this link is purely an argument of historical correlation. As such, correlation does not explain the serendipitous circumstances from 1983 to 1985 that culminated in Gorbachev's ascent to power. It does not explain why the Defense Intelligence Agency and CIA found Soviet military spending increasing slightly after 1985, when the U.S. defense budget was reaching its apex and would begin to decline in real terms. Eventual agreement to an increasingly irrelevant INF treaty is hardly evidence of Soviet collapse. According to Defense Secretary Cheney, Soviet strategic nuclear weapons modernization continued unabated as the new decade began. By contrast, the most dramatic unilateral Soviet concessions came in European conventional forces, where conservatives always complained of considerable Soviet superiority and where the Reagan buildup did almost nothing to change the balance.

Internal Soviet politics offers a far more plausible explanation. Gorbachev's actions were motivated in large measure by his own domestic political requirements and opportunities. Like American presidents, he has used foreign policy to achieve popularity and power at home. This power in turn helps him in his efforts to divert resources from the military to the economy, his most difficult problem. In using foreign policy to facilitate domestic goals, Gorbachev set an agenda to which the Reagan administration had to react. In a sense, Gorbachev outmaneuvered Reagan, just as the peace movement had briefly done several years earlier. As domestic pressure to end the Reagan buildup was mounting just before the revolutions of 1989, Gorbachev had myriad reasons to reach accommodation with the United States regardless of how much money Americans were spending on defense or what weapons they were building. External events do have an impact on the politics of defense, but we need to realize to what extent those events themselves are a product of domestic politics elsewhere in the world. Often we have less the clash of foreign policies than the international interaction of domestic policies.

While the relationship between the buildup and change in the communist bloc is difficult to judge, the relationship of the buildup to domestic politics is easy to discern. However incomplete or inconclusive, the Reagan revolution has changed the terrain of American politics. The 1980s witnessed the most significant restructuring of American politics since the New Deal, and defense policy shaped the struggle for control over this period of political change. Politicians and institutions used military policy to help create and respond creatively to the process of political change, to bend and bind social and international forces and events to their ends. The struggle involved conflict between and among some of the central institutions of American government and society over the course of several years. Of course American foreign and military policies had been used before for domestic political purposes, so the Reagan-era conflict represents a departure in degree rather than kind. In scope and duration it is unparalleled: Never before have specific military policies and defense spending played such an extensive and sustained role in domestic politics during *peacetime*.

The defense budgets and policies of the Reagan presidency must be viewed in the context of the broader Reagan legacy: the tax cuts, the deficit and the significantly increased interest payments on the federal debt, reductions in domestic programs, federal and Supreme Court appointments, and the general resurgence of conservatism. The new decade will test the durability of the change wrought in the Reagan era, including military policy and its effect on domestic politics; for the political impact of the Reagan buildup did not pass with the Reagan presidency or the end of the decade. President Reagan and Congress led the country onto a high plateau of defense spending from which descent is proving difficult if inevitable. The effects of the eight-year Reagan buildup, including significant responsibility for the deficit, massive interest on the debt, and deferred attention to social and economic problems, could constrain policy choices for years to come. The Reagan revolution exacerbated an already largely zero-sum reality of budgetary politics, a reality of painful opportunity costs. This political consequence will shape American politics regardless of the ideological inclinations of politicians, as George Bush discovered after less than a year in office. Whatever compromises Bush, Congress, and the public reach will be the influenced, if not determined, by political and fiscal constraints with which Reagan left us.

The peace dividend, unlike what the phrase implies, is not potential interest on an investment or a bonus for a job well done. Instead, we could measure the peace dividend in opportunity costs: what was expended in the military buildup during the last decade and how that expenditure and related policies have already constricted political choices at the beginning of the new decade. The extent to which the nation decides to devote military budget cuts to deflation of the deficit is one measure of the degree to which there is no peace dividend because it is already spent. The other measure is the extent to which the buildup has drained resources and attention from the social problems, in areas such as education, the environment, and the economy, which have become the focus of political pressure for governmental action. As the 1990s progress, these problems, the deficit, and the debate over the peace dividend are powerful reminders that the military buildup of the Reagan administration had less effect on the prospects for war and peace among nations than it has had and will continue to have on the American future.

Bibliography

Printed Primary Sources

American Friends Service Committee. *Call to Halt the Nuclear Arms Race: Proposal for a Mutual U.S.-Soviet Nuclear Weapons Freeze.* Philadelphia: AFSC, n.d. [1980].
Bruce, James T., Bruce W. MacDonald, and Ronald L. Tammen. *Star Wars at the Crossroads: The Strategic Defense Initiative after Five Years.* Staff report to Senators Bennett Johnston, Dale Bumpers, and William Proxmire, 12 June 1988.
Coalition for a Democratic Majority. *Democratic Solidarity: A Draft 1984 Party Platform on Foreign Policy and National Defense.* N.p., n.d. [c. June 1984].
Congressional Budget Office. *Budgeting for Defense Inflation.* January 1986.
———. *Strategic Defenses: Alternative Missions and Their Costs.* June 1989.
———. *An Analysis of the President's Budgetary Proposals for Fiscal Year 1992.* March 1991.
Congressional Record. 1980–89. Washington, D.C.
Council for a Livable World. *Draft Plank for Democratic Party Platform on Nuclear Arms Control—Preventing Nuclear War: A Strategy for Peace.* Boston: CLW, Spring 1984.
———. *Reagan Is Trying to Perpetuate His Military Policies.* Pamphlet. Boston: CLW, May 1987.
Defense Reorganization: Message from the President of the United States. Referred to the Senate Committee on Armed Services. 28 April 1986.
Deterrence and Survival in the Nuclear Age. Printed for the use of the Joint Committee on Defense Production of the U.S. Congress. Washington, D.C.: GPO, 1976.
General Accounting Office. *DOD Revolving Door: Post-DOD Employment May Raise Concerns.* GAO/NSIAD-87-116. April 1987.

Grassley, Charles, et al. *Should We Exempt Defense from Economic Reality? The Need for a Defense Budget Freeze.* Recommendation to the Senate and House Budget Committees and Republican Leadership. 7 February 1985.
Mayer, Andrew. *The Competition in Contracting Act: It's Application to the Department of Defense.* Congressional Research Service, Report No. 85-115F. 14 May 1985.
Moore, John Norton, and Robert F. Turner. *The Legal Structure of Defense Organization: Memorandum.* Prepared for the Presidential Blue Ribbon Commission on Defense Management. 15 January 1986.
National Conference of Bishops. *The Challenge of Peace: God's Promise and Our Response, a Pastoral Letter on War and Peace.* 3 May 1983.
National Party Platforms of 1980. Comp. Donald Bruce Johnson. Urbana: University of Illinois Press, 1982.
Official Proceedings of the 1984 Democratic National Convention. San Francisco, 16–19 July 1984.
Official Report of the Proceedings of the Democratic National Convention. New York, 11–14 August, 1980.
Official Report of the Proceedings of the Democratic National Convention and Committee, 1960. JFK Memorial Edition. Washington, D.C.: National Document Publishers, 1964.
President's Blue Ribbon Commission on Defense Management, David Packard, chairman. *An Interim Report to the President.* 28 February 1986.
———. *A Quest for Excellence: Final Report to the President.* June 1986.
———. *A Quest for Excellence: Appendix.* June 1986.
President's Commission on Strategic Forces. Brent Scowcroft, chairman. *Report of the President's Commission on Strategic Forces.* 6 April 1983.
The President's Strategic Defense Initiative. White House. January 1985.
Project on Military Procurement. *Who We Are, What We Do, Why We Do It.* N.d.
Pryor, David. "Defense Department's Inflation Windfall." *Congressional Record.* Vol. 131, no. 141. 22 October 1985.
Spinney, Franklin C. "Defense Facts of Life." Paper presented to the Subcommittee on Manpower and Personnel of the Senate Armed Services Committee, May 1980.
Taft, Robert, Jr. *White Paper on Defense: A Modern Military Strategy for the United States.* 1978 ed. Prepared with the assistance of William S. Lind. Published in cooperation with Senator Gary Hart. Washington, D.C., 15 May 1978.
U.S. Congress. House. Committee on Appropriations. Subcommittee on defense. *Department of Defense Appropriations for 1986.* P1. 99th Cong., 1st sess., 1985.
U.S. Congress. House. Committee on Armed Services. *Improved Conventional Force Capability: Raising the Nuclear Threshold.* Staff study prepared for the Research and Development Subcommittee and the Military Nuclear Systems Subcommittee of the House Armed Services Committee. 98th Cong., 1st sess., 1984.
U.S. Congress. House. Committee on Armed Services. *Report on the Duties and Authority of the Under Secretary of Defense (Acquisition).* 100th Cong., 1st sess., 16 November, 1987.
U.S. Congress. House. Committee on Armed Services. Subcommittee on Investigations. *Examination of Armed Services Policies and Procedures in the Procurement of*

Bibliography

Spare and Repair Parts, and the Pricing Thereof of These Items. 98th Cong., 2d sess., 19 and 20 April, 25 May, 9 June, 10 October, 1983.

U.S. Congress. House. Armed Services Committee. Investigations Subcommittee. *Hearings on HR 2545, Defense Procurement Reform Act of 1983*. 98th Cong., 1st sess., 27 April, 29 September, 19 October 1983.

U.S. Congress. House. Armed Services Committee. Investigations Subcommittee. *Hearings on HR 5064, Defense Spare Parts Procurement Reform Act and HR 4842*. 98th Cong., 2d sess., 13 and 21 March 1984.

U.S. Congress. House. Committee on Foreign Affairs. *Calling for a Mutual and Verifiable Freeze on and Reductions in Nuclear Weapons*. 98th Cong., 1st sess., 17 February, 2 and 8 March, 1983.

U.S. Congress. House. Committee on Foreign Affairs. *Review of Arms Control Implications of the Report of the President's Commission on Strategic Forces*. 98th Cong., 1st sess., 17, 19, 24 May 1983.

U.S. Congress. House. Committee on Foreign Affairs. Hearing before the Subcommittee on Arms Control, International Security, and Science. *ABM Treaty Interpretation Dispute*. 98th Cong., 2d sess., 22 October 1985.

U.S. Congress. House. Committee on Foreign Affairs. Subcommittee on International Security and Scientific Affairs. *Arms Control in Outer Space*. 98th Cong., 10 November 1983, 10 April, 2 May, 26 July 1984.

U.S. Congress. House. Committee on Government Operations. *Failure to Implement Effectively the Defense Department's High Dollar Spare Parts Breakout Program Is Costly*. Fifteenth Report by the Committee on Government Operations, 11 November 1983.

U.S. Congress. House. Committee on Government Operations. Subcommittee on Legislation and National Security. *Constitutionality of GAO's Bid Protest Function*. 99th Cong., 1st sess., 28 February, 7 March 1985, p. 1.

U.S. Congress. Senate. Committee on Armed Services. *Competition in Contracting Act of 1983*. 98th Cong., 1st sess., 7 and 8 June 1982.

U.S. Congress. Senate. Committee on Armed Services. *MX Missile Basing System and Related Issues*. 98th Cong., 1st sess., 18, 20, 21, 22, 26 April, 3 May 1983.

U.S. Congress. Senate. Committee on Armed Services. *Spare Parts Procurement for the Department of Defense*. 98th Cong., 1st sess., 26–27 October, 1983.

U.S. Congress. Senate. Committee on Armed Services. *Defense Organization: The Need for Change*. Staff Report. 16 October 1985.

U.S. Congress. Senate. Committee on Armed Services. *Reorganization of the Department of Defense*. 99th Cong., 1st sess., 16 October, 14, 19, 20, 21 November, 4, 5, 6, 11, 12 December 1985.

U.S. Congress. Senate. Committee on Armed Services. *Defense Acquisition Process*. 100th Congress, 2d sess., 11, 12, 27 July, 4 August 1988.

U.S. Congress. Senate. Committee on Armed Services. Subcommittee on Defense Acquisition Policy. *Implementation of the 1984 Defense Procurement Legislation*. 99th Cong., 1st sess., 10 and 17 October, 7 and 13 November 1985.

U.S. Congress. Senate. Committee on Foreign Relations. *Strategic Weapons Proposals: Part One*. 97th Cong., 1st sess., 1981.

U.S. Congress. Senate. Committee on Foreign Relations. *Nuclear Arms Reduction Proposals*. 97th Cong., 2d sess., 29 and 30 April, 11–13 May 1982.

U.S. Congress. Senate. Committee on Foreign Relations. *Controlling Space Weapons*. 98th Cong., lst. sess., 14 April, 18 May 1983.

U.S. Congress. Senate. Committee on Foreign Relations. *Strategic Defense and Anti-Satellite Weapons*. 98th Cong., 2d sess., 25 April 1984.

U.S. Department of Defense, Harold Brown, Secretary. *Report to the Congress on the FY 1981 Budget, FY 1982 Authorization Request and FY 1981–1985 Defense Programs*. 29 January 1980.

U.S. Department of Defense. *Report of the Secretary of Defense, Caspar W. Weinberger, to the Congress on the FY 1983 Budget, FY 1984 Authorization Request and FY 1983–1987 Defense Programs*. 8 February 1982.

U.S. Department of Defense. *Report of the Secretary of Defense Caspar W. Weinberger to the Congress on the FY 1984 Budget, the FY 1985 Authorization Request, and the FY 1984–1988 Defense Programs*. 1 February 1983.

U.S. Department of Defense. *Report of the Secretary of Defense Caspar W. Weinberger to the Congress on the FY 1986 Budget, FY 1987 Authorization Request, and FY 1986-1990 Defense Program*. 4 February 1985.

U.S. Department of Defense. *Report of the Secretary of Defense Caspar W. Weinberger to the Congress on the FY 1987 Budget, FY 1988 Authorization Request and FY 1987–1991 Defense Programs*. 5 February 1986.

U.S. Department of Defense. *Report of the Secretary of Defense Caspar W. Weinberger to the Congress on the FY 1988/FY 1989 Budget and FY 1988–1992 Defense Programs*. 12 January 1987.

U.S. Department of Defense, Dick Cheney, Secretary. *Defense Management: Report to the President*. July 1989.

U.S. Department of Defense. Directorate for Information Operations and Reports. *100 Companies Receiving the Largest Dollar Volume of Military Prime Contract Awards*. Issued annually, fiscal years 1979–85.

U.S. Federal Election Commission. *FEC Reports on Financial Activity, 1980–81: Final Report, Party and Non-Party Political Committees, Vol. IV*.

U.S. Federal Election Commission. *FEC Reports on Financial Activity, 1982–83: Final Report, Party and Non-Party Political Committees, Vol. IV*.

U.S. Federal Election Commission. *FEC Reports on Financial Activity, 1984–85: Final Report, Party and Non-Party Political Committees, Vol. IV*.

Books

Adams, Gordon. *The Politics of Defense Contracting: The Iron Triangle*. New Brunswick, N.J.: Transaction Books, 1982.

——. *Controlling Weapons Costs: Can the Pentagon Reforms Work?*. New York: Council on Economic Priorities, 1983.

Aliano, Richard. *American Defense Policy from Eisenhower to Kennedy: The Politics of Changing Military Requirements, 1957–1961*. Athens: Ohio University Press, 1975.

Allison, Graham. *The Essence of Decision*. Boston: Little, Brown, 1971.

Almond, Gabriel. *The American People and Foreign Policy*. New York: Praeger, 1960.

Anderson, Martin. *Revolution*. New York: Harcourt Brace Jovanovich, 1988.

Ball, Desmond. *Politics and Force Levels: The Strategic Missile Program of the Kennedy Administration*. Berkeley: University of California Press, 1980.

Bibliography

Barrett, Archie D. *Reappraising Defense Organization*. Washington, D.C.: National Defense University Press, 1983.
Boston Study Group. *Winding Down: The Price of Defense*. San Francisco: W. H. Freeman, 1982.
Bruce-Briggs, B., ed. *The New Class?* New Brunswick, N.J.: Transaction Books, 1979.
Burkhart, James A., and Frank J. Kendrick, eds. *The New Politics: A Mood or Movement?*. Englewood Cliffs, N.J.: Prentice-Hall, 1971.
Burnham, Walter Dean. *The Current Crisis in American Politics*. New York: Oxford University Press, 1982.
Center on Budget and Policy Priorities. *The Fiscal 1986 Defense Budget: The Weapons Buildup Continues*. Washington, D.C.: Center on Budget and Policy Priorities, 1985.
Cincinnatus. *Self-Destruction: The Disintegration and Decay of the U.S. Army during the Vietnam Era*. New York: Norton, 1981.
Clark, Asa A., IV, et al., eds. *The Defense Reform Debate: Issues and Analysis*. Baltimore: Johns Hopkins University Press, 1984.
Clubb, Jerome, William Flanigan, and Nancy Zingale. *Partisan Realignment: Voters, Parties, and Government in American History*. Beverly Hills, Calif.: Sage, 1980.
Coates, James, and Michael Killian. *Heavy Losses: The Dangerous Decline of American Defense*. New York, Viking: 1985.
Codevilla, Angelo. *While Others Build: The Commonsense Approach to the Strategic Defense Initiative*. New York: Free Press, 1988.
Cohen, Bernard C. *The Public's Impact on Foreign Policy*. Boston: Little, Brown, 1973.
Crawford, Alan. *Thunder on the Right*. New York: Pantheon, 1980.
Crotty, William J., and Gary C. Jacobson. *American Parties in Decline*. Boston: Little, Brown, 1980.
Davis, Larry J. *The Emerging Democratic Majority: Lessons and Legacies from the New Politics*. New York: Stein and Day, 1974.
De Benedetti, Charles. *The Peace Reform in American History*. Bloomington,: Indiana University Press, 1980.
Divine, Robert A. *Foreign Policy and U.S. Presidential Elections*. New York: Watts, New Viewpoints, 1974.
——. *Blowing on the Wind: the Nuclear Test Ban Debate, 1954–1960*. New York: Oxford University Press, 1978.
Drell, Sidney, et al., eds. *The Reagan Strategic Defense Initiative: A Technical, Political, and Arms Control Assessment*. Cambridge, Mass: Ballinger, 1985.
Edsall, Thomas Byrne. *The New Politics of Inequality*. New York: Norton, 1984.
Edwards, John. *Superweapon: The Making of MX*. New York: Norton, 1982.
Enthoven, Alain C., and K. Wayne Smith. *How Much Is Enough?* New York: Harper and Row, 1971.
Evangelista, Matthew. *Innovation and the Arms Race: How the United States and the Soviet Union Develop New Military Technologies*. Ithaca, N.Y.: Cornell University Press, 1988.
Evans, Peter, et al., eds. *Bringing the State Back In*. New York: Cambridge University Press, 1985.
Fallows, James. *National Defense*. New York: Random House, 1981.

Fine, Melinda, and Peter M. Steven, eds. *American Peace Directory.* Brookline, Mass.: Institute for Defense and Disarmament Studies, 1984.
Fiorina, Morris. *Congress: Keystone of the Washington Establishment.* New Haven, Conn.: Yale University Press, 1977.
Fitzgerald, A. Ernest. *The Pentagonists.* Boston: Houghton, Mifflin, 1989.
Forum Institute. *1983 Handbook: Arms Control and Peace Organizations/Activities.* Washington, D.C.: Forum Institute, 1983.
Freeman, Jo, ed. *Social Movements of the Sixties and Seventies.* New York: Longman, 1983.
Gabriel, Richard A. *Military Incompetence.* New York: Hill and Wang, 1985.
Gabriel, Richard A., and Paul Savage. *Crisis in Command: Mismanagement in the Army.* New York: Hill and Wang, 1978.
Gaddis, John Lewis. *Strategies of Containment.* New York: Oxford University Press, 1983.
George, Alexander. *Presidential Decisionmaking in Foreign Policy.* Boulder, Colo.: Westview, 1980.
Ginsberg, Benjamin, and Martin Shefter. *Politics by Other Means: Institutional Conflict and the Declining Significance of Elections in America.* New York: Basic Books, 1990.
Goodwin, Jacob. *Brotherhood of Arms: General Dynamics and the Business of Defending America.* New York: Times Books, 1985.
Graham, Daniel O. *The Non-Nuclear Defense of Cities: The High Frontier Space-Based Defense against ICBM Attack.* Cambridge, Mass.: Abt Books, 1983.
———. *To Provide for the Common Defense: The Case for Space Defense.* Copyright by Daniel O. Graham, 1986.
Hadley, Arthur T. *The Straw Giant—Triumph and Failure: America's Armed Forces.* New York: Random House, 1984.
Haffa, Robert P. *The Half War: Planning U.S. Rapid Deployment Forces to Meet a Limited Contingency, 1960–1983.* Boulder, Colo.: Westview, 1984.
Halloran, Richard. *To Arm a Nation: Rebuilding America's Endangered Defenses.* New York: Macmillian, 1986.
Halperin, Morton. *Bureaucratic Politics and Foreign Policy.* Washington, D.C.: Brookings, 1974.
Halstead, Fred. *Out Now! A Participant's Account of the American Movement against the Vietnam War.* New York: Monad, 1978.
Hart, Gary, and William S. Lind. *America Can Win: The Case for Military Reform.* Bethesda, Md.: Adler and Adler, 1986.
Hartung, William D., et al. *The Strategic Defense Initiative: Costs, Contracts, and Consequences.* New York: Council on Economic Priorities, 1985.
Holland, Lauren H., and Robert A. Hoover. *The MX Decision: A New Direction in U.S. Weapons Procurement Policy?* Boulder, Colo.: Westview, 1985.
Huntington, Samuel P. *The Common Defense.* New York: Columbia University Press, 1961.
Inglehart, Ronald. *The Silent Revolution: Changing Styles among Western Publics.* Princeton, N.J.: Princeton University Press, 1977.
Institute for Defense and Disarmament Studies. *Peace Resource Book: A Comprehensive Guide to Issues, Groups, and Literature.* Cambridge, Mass.: Ballinger, 1986.
Janis, Irving. *Victims of Groupthink.* Boston: Houghton Mifflin, 1972.

Jervis, Robert. *Perception and Misperception in International Politics*. Princeton, N.J.: Princeton University Press, 1979.
Katz, Milton S. *Ban the Bomb: A History of SANE, 1957–1985*. New York: Greenwood, 1985.
Kayden, Xandra, and Eddie Mahe. *The Party Goes On: The Persistence of the Two-Party System*. New York: Basic Books, 1985.
Kennedy, Edward M., and Mark O. Hatfield. *Freeze! How You Can Help Prevent Nuclear War*. New York: Bantam, 1982.
Kotz, Nick. *Wild Blue Yonder: Money, Politics, and the B-1 Bomber*. New York: Pantheon, 1988.
Krulak, Victor H. *Organization for National Security*. Washington, D.C.: U.S. Strategic Institute, 1983.
Lebow, Richard Ned. *Between Peace and War*. Baltimore: Johns Hopkins University Press, 1981.
Lowi, Theodore J. *The Personal President*. Ithaca, N.Y.: Cornell University Press, 1985.
Luttwak, Edward. *The Pentagon and the Art of War: The Question of Military Reform*. New York: Simon and Schuster, 1984.
McCrea, Frances B., and Gerald E. Markle. *Minutes to Midnight: Nuclear Weapons Protest in America*. Newbury Park, Calif.: Sage, 1989.
MacDougall, Walter. *The Heavens and the Earth*. New York: Basic Books, 1985.
Magraw, Katherine. "SDI: Fading Fantasy or Fait Accompli?" Washington, D.C.: Spacewatch, 1988.
May, Ernest R. *Lessons of the Past*. New York: Oxford, 1973.
Mayer, John D. *Rapid Deployment Forces: Policy and Budgetary Implications*. Washington, D.C.: Congressional Budget Office, February 1983.
Mayhew, David. *The Electoral Connection*. New Haven, Conn.: Yale University Press, 1974.
Melman, Seymour. *Pentagon Capitalism*. New York: McGraw-Hill, 1970.
Meyer, David S. *A Winter of Discontent: The Nuclear Freeze and American Politics*. New York: Praeger, 1990.
Miller, Steven E. and Stephen Van Evera, eds. *The Star Wars Controversy: An International Security Reader*. Princeton, N.J.: Princeton University Press, 1986.
Miller, Warren E., and Teresa Levitan. *Leadership and Change: The New Politics and the American Electorate*. Cambridge, Mass.: Winthrop, 1976.
Murphy, Thomas P. *The New Politics Congress*. Lexington, Mass.: Lexington Books, 1974.
Nelson, Michael, ed. *The Elections of 1984*. Washington, D.C.: Congressional Quarterly, 1985.
——. *The Presidency and the Political System*. 2d ed. Washington, D.C.: Congressional Quarterly, 1988.
Nolan, Janne. *Guardians of the Arsenal: The Politics of Nuclear Strategy*. New York: Basic Books, 1989.
Nordlinger, Eric. *On the Autonomy of the Democratic State*. Cambridge, Mass.: Harvard University Press, 1981.
Petrocik, John. *Party Coalitions: Realignment and the Decline of the New Deal Party System*. Chicago: University of Chicago Press, 1981.
Polsby, Nelson W. *Consequences of Party Reform*. New York: Oxford University Press, 1983.

Posen, Barry R. *The Sources of Military Doctrine*. Ithaca, N.Y.: Cornell University Press, 1984.
Pressler, Larry. *Star Wars: The Strategic Defense Debates in Congress*. New York: Praeger, 1986.
Rasor, Dina. *The Pentagon Underground*. New York: Times Books, 1985.
———, ed. *More Bucks, Less Bang: How the Pentagon Buys Ineffective Weapons*. Washington, D.C.: Fund for a Constitutional Government, 1983.
Rearden, Steven L. *The Evolution of American Strategic Doctrine*. Boulder, Colo.: Westview, 1984.
Rockefeller Brothers Fund. *International Security: The Military Aspect*. Garden City, N.Y.: RBF, 1958.
Romjue, John L. *From Active Defense to AirLand Battle: The Development of Army Doctrine, 1973–1982*. Fort Monroe, Va.: U.S. Army Training and Doctrine Command, June 1984.
Rosen, Steven, ed. *Testing the Theory of the Military Industrial Complex*. Lexington, Mass.: Heath, 1973.
Rosenau, James, ed. *The Domestic Sources of Foreign Policy*. New York: Free Press, 1967.
Russett, Bruce. *What Price Vigilance?* New Haven, Conn.: Yale University Press, 1970.
Sanders, Jerry W. *Peddlers of Crisis: The Committee on the Present Danger and the Politics of Containment*. Boston: South End, 1983.
Sarkesian, Sam, ed. *The Military Industrial Complex: A Reassessment*. Beverly Hills, Calif.: Sage, 1972.
Schattschneider, E. E. *The Semisovereign People: A Realist's View of Democracy in America*. Hinsdale, Il.: Dryden, 1975.
Scheer, Robert. *With Enough Shovels: Reagan, Bush, and Nuclear War*. New York: Random House, Vintage, 1983.
Schlesinger, Arthur M., Jr. *The Imperial Presidency*. Boston: Houghton Mifflin, 1973.
Schumpeter, Joseph. *Capitalism, Socialism, and Democracy*. New York: Harper, 1942.
Schurmann, Franz. *The Logic of World Power*. New York: Pantheon, 1974.
Scoville, Herbert. *MX: Prescription for Disaster*. Cambridge, Mass.: MIT Press, 1981.
Shafer, Byron E. *Quiet Revolution: The Struggle for the Democratic Party and the Shaping of Post-Reform Politics*. New York: Russell Sage, 1983.
Smith, Hedrick. *The Power Game: How Washington Really Works*. New York: Random House, 1988.
Spinney, Franklin C. *Defense Facts of Life: The Plans/Reality Mismatch*. Ed. James Clay Thompson. Boulder, Colo.: Westview, 1985.
Stares, Paul. *Space Weapons and U.S. Strategy: Origins and Development*. London: Croom Helm, 1985.
Steinberg, Gerald, ed. *Lost in Space: The Domestic Politics of the Strategic Defense Initiative*. Lexington, Mass: Lexington Books, 1988.
Stockman, David. *The Triumph of Politics*. New York: Avon, 1987.
Stubbing, Richard. *The Defense Game*. New York: Harper and Row, 1986.
Summers, Harry G., Jr. *On Strategy: A Critical Analysis of the Vietnam War*. Novato, Calif.: Presidio, 1982.

Sundquist, James. *The Decline and Resurgence of Congress.* Washington, D.C.: Brookings, 1981.
Talbott, Strobe. *Deadly Gambits.* New York: Knopf, 1985.
U.S. Defense Policy. Washington, D.C.: Congressional Quarterly, 1983.
Viguerie, Richard A. *The New Right: We're Ready to Lead.* Falls Church, Va.: Viguerie Co., 1981.
Waller, Douglas C. *Congress and the Nuclear Freeze.* Amherst: University of Massachusetts Press, 1987.
Waltz, Kenneth. *Foreign Policy and Democratic Politics.* Boston: Little, Brown, 1967.
——. *Theory of International Politics.* Reading, Mass.: Addison-Wesley, 1979.
Wattenberg, Martin P. *The Decline of American Political Parties, 1952–1980.* Boston: Harvard, 1984.
Weiner, Tim. *Blank Check: The Pentagon's Black Budget.* New York: Warner, 1990.
White, Theodore H. *The Making of the President, 1960.* New York: Atheneum, 1961.
Wittner, Lawrence S. *Rebels against War: The American Peace Movement, 1941–1960.* New York: Columbia University Press, 1969.
Wolfe, Alan. *The Rise and Fall of the Soviet Threat: Domestic Sources of the Cold War Consensus.* Boston: South End, 1984.
Yarmolinsky, Adam. *The Military Establishment.* New York: Harper and Row, 1971.
Yarnella, Ernest. *The Missile Defense Controversy.* Lexington: University Press of Kentucky, 1977.
Zaroulis, Nancy, and Gerald Sullivan. *Who Spoke Up? American Protest Against the War in Vietnam, 1963-1975.* New York: Holt, Rinehart and Winston, 1984.

Articles

Aspin, Les. "The Defense Budget and Foreign Policy: The Role of Congress." *Daedalus* 104 (Summer 1975): 155–74.
Axelrod, Robert. "Where the Votes Come From: An Analysis of Electoral Coalitions, 1952–1968." *American Political Science Review* 66 (1972): 11–20.
——. "Presidential Election Coalitions in 1984." *American Political Science Review* 80 (1986): 281–84.
Ball, George W. "The War for Star Wars." *New York Times Review of Books*, 11 April 1985, pp. 38–44.
Beck, Robert J. "Munich's Lessons Reconsidered." *International Security* 14 (Fall 1989): 160–191.
Bennett, Charles. "The Rush to Deploy SDI." *Atlantic Monthly*, April 1988, pp. 53–61.
Bernstein, Robert A., and William W. Anthony. "The ABM Issue in the Senate, 1968–1970: The Importance of Ideology." *American Political Science Review* 68 (1974): 1198–1206.
Boyer, Paul. "From Activism to Apathy: The American People and Nuclear Weapons, 1963–1980." *Journal of American History* 70 (1984): 821–44.
Brint, Stephen. "'New Class' and Cumulative Trend Explanations of the Liberal Political Attitudes of Professionals." *American Journal of Sociology* 10 (1984): 30–71.

Bundy, McGeorge, et al. "Nuclear Weapons and the Atlantic Alliance." *Foreign Affairs* 60 (Spring 1982): 753–68.
Canby, Steven L. "U.S. Defense Policy: The problem Is Not More Money." *AEI Foreign Policy and Defense Review* 1, no. 3 (1979): 23–36.
Cox, Arthur Macy. "The CIA's Tragic Error." *New York Review of Books*, 6 November 1980, pp. 21–24.
Cummings, Bruce. "Chinatown: Foreign Policy and Elite Realignment." In *The Hidden Election*, ed. Thomas Ferguson and Joel Rogers, pp. 196–231. New York: Pantheon, 1981.
Davidon, Ann Morrissett. "The U.S. Anti-Nuclear Movement." *Bulletin of the Atomic Scientists*, December 1979, pp. 45–48.
Deitchman, Seymour J. "Weapons, Platforms, and the New Armed Services." *Issues in Science and Technology* 1 (Spring 1985): 83–99.
Drew, Elizabeth. "A Political Journal." *New Yorker*, 20 June 1983, pp. 39–75.
Easterbrook, Gregg. "DIVAD." *Atlantic Monthly*, October 1982, pp. 29–39.
———. "Why DIVAD Wouldn't Die." *Washington Monthly*, November 1984, pp. 10–22.
Ehrenreich, Barbara, and John Ehrenreich. "The Professional-Managerial Class." In *Between Labor and Capital*, ed. Pat Walker, pp. 4–45. Boston: South End, 1979.
Ellsberg, Daniel. "The Quagmire Myth and the Stalemate Machine." In *Papers on the War*, pp. 42–135. New York: Simon and Schuster, 1972.
Evangelista, Matthew. "Issue-Area and Foreign Policy Revisited." *International Organization* 43 (Winter 1989): 149–71.
Fallows, James. "The Spend-Up." *Atlantic Monthly*, July 1986, pp. 27–31.
Feighan, Edward. "The Freeze in Congress." In *The Nuclear Freeze Debate*, ed. Paul M. Cole and William J. Taylor, Jr., pp. 29–55. Boulder, Colo.: Westview, 1983.
Gansler, Jacques S. "Defense: A Demonstration Case for Industrial Strategy." *Challenge* 26 (January/February 1984): 58–61.
Gelb, Leslie H. "Aides Urged Reagan to Postpone Antimissile Ideas for More Study." *New York Times*, 25 March 1983, pp. A1, 8.
Ginsberg, Benjamin, and Martin Shefter. "The Setting: A Critical Realignment?" In *The Elections of 1984*, ed. Michael Nelson, pp. 1–26. Washington, D.C.: Congressional Quarterly, 1985.
Gourevitch, Peter. "The Second Image Reversed: The International Sources of Domestic Politics." *International Organization* 32 (Autumn 1978): 881–912.
Graham, Daniel O. "Toward a New U.S. Strategy: Bold Strokes Rather than Increments." *Strategic Review* 9 (Spring 1981): 9–16.
Graham, Thomas W., and Bernard M. Kramer. "The Polls: ABM and Star Wars—Attitudes toward Nuclear Defense, 1945–1985." *Public Opinion Quarterly* 50 (Winter 1986): 125–34.
Gray, Colin S. "A New Debate on Ballistic Missile Defense." *Survival* 23 (March/April 1981): 60–71.
Gray, Colin S., and Keith Payne, "Victory Is Possible." *Foreign Policy* (Summer 1980): 14–27.
Gray, Robert C. "Arms Control Implications of Ballistic Missile Defense." In *Antiballistic Missile Defense in the 1980s*, ed. Ian Bellamy and Coit D. Blacker, pp. 29–45. London: Cass, 1983.

Bibliography

Halperin, Morton. "The Gaither Committee and the Policy Process." *World Politics* 13 (April 1961): 360–84.
Hammond, Paul Y. "NSC-68: Prologue to Rearmament." In *Strategy, Politics, and Defense Budgets*, ed. Warner R. Schilling, Paul Y. Hammond, and Glenn H. Synder, pp. 271–378. New York: Columbia University Press, 1962.
Hart, Gary. "The Case for Military Reform." *Wall Street Journal*, 23 January 1981, p. 20.
———. "What's Wrong with the Military." *New York Times Magazine*, 14 February 1982, pp. 16–19, 40, 41, and 45.
Herken, Gregg. "The Earthly Origins of Star Wars." *Bulletin of the Atomic Scientists*, October 1987, pp. 20-28.
Hess, Stephen, and Michael Nelson. "Foreign Policy: Dominance and Decisiveness in Presidential Elections." In *The Elections of 1984*, ed. Michael Nelson, pp. 129–54. Washington, D.C.: Congressional Quarterly, 1985.
Hoffman, David. "Reagan Seized Idea Shelved in 1980 Race." *Washington Post*, 3 March 1985, pp. A1, 18.
Hoffman, David, and Lou Cannon. "President Overruled Advisors on Announcing Defense Plans." *Washington Post*, 26 March 1983, pp. A1, 7.
Huntington, Samuel P. "Congressional Responses to the Twentieth Century." In *The Congress and America's Future*, ed. David Truman, pp. 6–38. Englewood Cliffs, N.J.: Prentice-Hall, 1965.
Jones, David C. "What's Wrong with Our Defense Establishment?" *New York Times Magazine*, 7 November 1982, pp. 38–42, 70, 83.
Kaplan, Fred. "Return of the ABM." *Atlantic Monthly*, September 1981, pp. 8–18.
Keeter, Scott. "Public Opinion in 1984." In *The Election of 1984*, ed. Gerald Pomper et al., pp. 91–111. Chatham, N.J.: Chatham House, 1985.
Kemp, Jack F. "U.S. Strategic Force Modernization: A New Role for Missile Defense?" *Strategic Review* 7 (Summer 1980): 11–17.
Kissinger, Henry. "Domestic Structures and Foreign Policy." *Daedalus* 95 (Spring 1966): 503–29.
Klare, Michael T. "Army in Search of a War." *Progressive*, February 1981, pp. 18–23.
Krasner, Stephen. "Are Bureaucracies Important?" *Foreign Policy* (Summer 1972): 159–79.
Kurth, James. "Why We Buy the Weapons We Do." *Foreign Policy* (Summer 1973): 33–56.
Ladd, Everett Carll. "The Shifting Party Coalitions: From the 1930s to the 1970s." In *Party Coalitions in the 1980s*, ed. Seymour Martin Lipset, pp. 127–49. San Francisco: Institute for Contemporary Studies, 1981.
Lehman, John. "The Six Hundred–Ship Navy: Why We Need It and How We Can Afford It." *Defense* (January/February 1986): 14–21.
Lindsay, James M. "Congress and Defense Policy: 1961 to 1986." *Armed Forces and Society* 13 (Spring 1987): 371–400.
Lodal, Jan M. "Assuring Strategic Stability: An Alternative View." *Foreign Affairs* 54 (April 1976): 462–81.
Luttwak, Edward. "The American Style of Warfare and the Military Balance." *Survival* 21 (March/April 1979): 57.
McCormick, James M. "Congressional Voting on the Nuclear Freeze Resolutions." *American Politics Quarterly* 13 (January 1985): 122-36.

McNaugher, Thomas. "Weapons Procurement: The Futility of Reform." *International Security* 12 (Fall 1987): 63–104.
McWilliams, Carey. "Time for a New Politics." *Nation*, 26 May 1962, p. 466.
Markusen, Ann. "The Militarized Economy." *World Policy Journal* 3 (Summer 1986): 495–516.
Meyer, Edward C. "The JCS: How Much Reform Is Needed?" *Armed Forces Journal* 119 (April 1982): 82–90.
Mills, Derek M. "From New Politics to Mass Catharsis." *War/Peace Report*, October 1967, pp. 8–9.
Morrison, David C. "Caucusing for Reform." *National Journal*, 28 June 1986, pp. 1596–1602.
Mustin, H. C. "The Role of the Navy and Marines in the Norwegian Sea." *Naval War College Review* 39 (March/April 1986): 2–6.
Nitze, Paul. "Assuring Strategic Stability in an Era of Detente." *Foreign Affairs* 54 (January 1976): 207–32.
———. "Deterring Our Deterrent." *Foreign Policy* (Winter 1976–77): 195–210.
Nye, Joseph S. and Sean M. Lynn-Jones. "International Security Studies: A Report of a Conference on the State of the Field." *International Security* 12 (Spring 1988): 5–27.
Owens, MacKubin Thomas. "The Hollow Promise of JCS Reform." *International Security*. 10 (Winter 1985–86): 69–97.
———. "American Strategic Culture and Civil-Military Relations: The Case of JCS Reform." *Naval War College Review* 39 (March/April 1986): 43–59.
Paine, Christopher. "Nuclear Combat: The Five-Year Defense Plan." *Bulletin of the Atomic Scientists*, November 1982, pp. 5–12.
Pipes, Richard. "Why the Soviet Union Thinks It Could Fight and Win a Nuclear War." *Commentary*, July 1977, pp. 21–34.
Quirt, John. "Washington's New Push for Anti-Missiles." *Fortune*, 19 October 1981, pp. 142–48.
Record, Jeffrey. "The Military Reform Caucus." *Washington Quarterly* 6 (Spring 1983): 125–29.
Reich, Robert. "High Tech, a Subsidiary of Pentagon Inc." *New York Times*, 29 May 1985, p. A23.
———. "Reagan's Hidden 'Industrial Policy.'" *New York Times*, 4 August 1985, sec. 3, p. 3.
Reichard, Gary W. "The Domestic Politics of National Security." In *The National Security: Its Theory and Practice, 1945–1960*, ed. Norman A. Graebner, pp. 243–74. New York: Oxford University Press, 1986.
Richelson, Jeffrey. "PD-59, NSDD-13 and the Reagan Strategic Modernization Program." *Journal of Strategic Studies* 6 (June 1983): 128–46.
Rosen, Stephen Peter. "Nuclear Arms and Strategic Defense." *Washington Quarterly* 4 (Spring 1981): 82–99.
Shapley, Deborah. "The Army's New Fighting Doctrine." *New York Times Magazine*, 28 November 1982, pp. 36–42+.
Shefter, Martin. "Party, Bureaucracy, and Political Change in the United States." In *Political Parties: Development and Decay*, ed. Louis Maisel and Joseph Cooper, pp. 211–66. Beverly Hills, Calif.: Sage, 1978.
Skocpol, Theda. "Political Response to Capitalist Crisis: NeoMarxist Theories of the State and the Case of the New Deal." *Politics and Society* 10 (1980): 155–201.

Smith, R. Jeffrey. "Reagan Plans New ABM Effort." *Science*, 8 April 1983, pp. 170–71.
Sprey, Pierre. "Mach 2: Reality or Myth?" *International Defense Review* 8 (1980): 1209–12.
Sprey, Pierre, and Jack Merrit. "Quality, Quantity or Training." *U.S.A.F. Fighter Weapons Review*, 27 (Summer 1979): 9.
Stanley, Harold W., William T. Bianco, and Richard G. Niemi. "Partisanship and Group Support over Time: A Multivariate Analysis." *American Political Science Review* 80 (1986): 969–76.
Starry, Donn. "Extending the Battlefield." *Military Review* 61 (March 1981): 31–50.
Stein, Arthur. "Strategy as Politics, Politics as Strategy: Domestic Debates, Statecraft, and Star Wars." In *The Logic of Nuclear Terror*, ed. Roman Kolkowicz, pp. 186–210. Boston: Allen and Unwin, 1987.
Taylor, Maxwell D. "Changing Military Priorities." *AEI Foreign Policy and Defense Review* 1, no. 3, (1980): 2–13.
Turner, Stansfield. "Toward a New Defense Strategy." *New York Times Magazine*, 10 May 1981, pp. 15–17.
Vandercook, William F. "SDI Show Hits the Road." *Bulletin of the Atomic Scientists*, October 1986, pp. 16–18.
Vogel, David. "The Public Interest Movement and the American Reform Tradition." *Political Science Quarterly* 95 (Winter 1980–81): 607–27.
———. "The Power of Business in America: A Reappraisal." *British Journal of Political Science* 13 (January 1983): 19–44.
Wallop, Malcolm. "Opportunities and Imperatives of Ballistic Missile Defense." *Strategic Review* 7 (Fall 1979): 13–21.
Weinberger, Caspar. "U.S. Defense Strategy." *Foreign Affairs* 64 (Spring 1986): 675–97.
Wells, Samuel F. "Sounding the Tocsin: NSC 68 and the Soviet Threat," *International Security* 4 (Fall 1979): 116–58.
Wieseltier, Leon. "The Pitfalls of the Peace Issue." *Washington Post*, 27 May 1984, p. C5.
"The Winds of Reform." *Time*, 7 March 1983, pp. 12–30.
Wirls, Daniel. "Reinterpreting the Gender Gap." *Public Opinion Quarterly* 50 (Fall 1986): 316–30.
Wirthlin, Richard. "The Republican Strategy and Its Electoral Consequences." In *Party Coalitions in the 1980s*, ed. Seymour Martin Lipset, pp. 235–66. San Francisco: Institute for Contemporary Studies, 1981.
Wohlstetter, Albert. "Optimal Ways to Confuse Ourselves." *Foreign Policy* (Fall 1975): 170–98.
Wood, Robert S., and John Hanley, Jr. "The Maritime Role in the North Atlantic." *Naval War College Review* 38 (November/December 1985): 5–18.
Yankelovich, Daniel, and John Doble. "The Public Mood: Nuclear Weapons and the USSR." *Foreign Affairs* 63 (Fall 1984): 33–46.
Yankelovich, Daniel, and Larry Kaagan. "Assertive America." *Foreign Affairs: America and the World, 1980* 59 (1981): 696–713.

Index

Abrahamson, James A., 156, 161
Adelman, Kenneth, 117
Afghanistan: and SALT II, 27
AirLand battle plan, 83
ALCM. *See* cruise missile
Aliano, Richard, 5n, 6n
Allen, Richard, 144, 146
American Conservative Union, 25, 105
American Friends Service Committee, 66n, 67n, 71, 72–73
American Security Council, 25
Anderson, John, 30
Anderson, Martin, 140, 144–46, 147
Andrews, Mark, 92
antiballistic missiles (ABM), 152–53; treaty, 70, 137–38, 164
antimilitarism, 62–66
antisatellite weapons (ASAT), 136–38, 141. *See also* antiballistic missiles: treaty; ballistic missile defense
antiwar movement: and nuclear weapons peace movement, 58–61
Arms Control and Disarmament Agency (ACDA), 25, 58, 117
Arms Export Control Act, 61
ASAT. *See* antisatellite weapons
Askew, Reuben, 122
Aspin, Les, 112n, 197n; and Democratic Policy Commission, 130n; and MX missile, 115–17; on SDI, 159
AuCoin, Les, 221

Babbitt, Bruce, 130
Ball, Desmond, 6n
Ballistic Missile Boost Intercepts (Bambi), 137
ballistic missile defense (BMD), 136–38, 141–42; and SDI, 145. *See also* antiballistic missiles: treaty; antisatellite weapons
Barnes, Mike, 212
Barron, John, 105–6
base closures, 212–13
Bendetsen, Karl, 144–45
Biden, Joseph, 130
bilateral arms control, 69–78. *See also* freeze movement
BMD. *See* ballistic missile defense
Boeing, 205
B-1 bomber: cancellation of, 20–21, 64; peace movement opposition to, 117–18; and procurement reform, 169
Boxer, Barbara, 213
Boyd, John, 82–83
Broomfield, William, 104n
Brown, Harold, 65, 114n, 115
build-down, 118; and MX missile, 116–17; and SDI, 158
Bumpers, Dale, 106
Bundy, McGeorge, 69
Bush, George, 22, 199; and 1984 campaign, 128; and 1988 campaign, 210–

Bush, George (cont.)
11; and nuclear forces buildup, 37; and peace dividend, 221
Bush administration, 117; and military budget, 212–23
business community: and military Keynesianism, 50–52

Carlucci, Frank, 212–13
Carnegie Foundation, 74
Carter, Jimmy, 21, 29; attempts to limit military expenditures, 32; election of, 20
Carter administration: and cold war Democrats, 22; and conservative advocacy groups, 25; and New Politics, 64–65
Catholic bishops: endorsement of peace movement, 120
Center for Defense Information, 66n
Central Intelligence Agency (CIA), 22–23, 61, 80
Cheney, Richard: and military budget, 215–20; and Soviet military capability, 224–25
Chrysler Corporation, 51–52
civil rights movement: coalition with peace movement, 58–61
Clark, William: and SDI, 145–46, 149–51
Clausewitz, Karl von, 82–83
Clergy and Laity Concerned, 71
Coalition for a Democratic Majority, 129
Coalition on National Priorities and Military Policy, 63
Coalition for a New Foreign Policy, 63
Coalition for a New Foreign and Military Policy, 63, 66n, 67n
Codevilla, Angelo, 139
Coffin, William Sloane, 201
Cohen, William: and build-down, 116
cold war, 1–2; and dealignment, 17–21; and militarism, 66; and 1980 elections, 28–30; politics of, 11–17; and Republican party, 21–28
Committee for Non-Violent Action, 58
Committee on the Present Danger (CPD), 22–23
Committee for the Survival of a Free Congress, 22
Common Cause, 66n, 202n
Congress, U.S.: institutional base of military reform, 4, 79–81; and MX missile, 113–18; and nuclear freeze resolution, 102–5, 110–12; and Reagan administration clash with peace movement, 90–91; and SDI, 159–60. *See also* Military Reform Caucus; *names of specific committees*
Conservative Caucus, 22; and SDI, 146
containment, 66
Conte, Silvio, 110
conventional deterrence, 68
conventional forces and strategy: "attrition" warfare, 82; "maneuver" warfare, 82–83; Reagan buildup, 40–43
Coors, Joseph, 144–45
Corn, David, 105n, 109n
Council on Economic Priorities, 95n
Council for a Livable World, 66n, 67n, 73, 123; congressional procurement reform, 169
Courter, Jim, 91n, 92, 171; and procurement reform, 172
Cranston, Alan, 92, 122, 123, 221
Crowe, William, 219
cruise missile, 65, 68, 199

Darman, Richard, 214, 219
dealignment, political, 17–21
Deaver, Michael, 140
Defense Acquisitions Board, 81
Defense, Department of (DOD): and DIVAD, 174; employment, 48; implementation of SDI, 154; and industrial policy, 49–50; and military reform, 4, 87–88; and operational testing, 172–73; and SDI, 141–44
defense policy, 4–5, 7–8, 31; and Mondale campaign, 125–28; and 1984 Reagan campaign, 127–28; and Republican coalition building, 1
Defensive Technologies Study Team: and SDI, 155
DeLauer, Richard: on ABM program, 143; and critique of military reform, 171; and Military Reform Caucus, 169
Democratic Leadership Council (DLC), 129–30
Democratic party: dealignment, 203; delegate selection rules, 61; dependence on New Politics coalition, 62; and Gaither report, 14–15; and military buildup, 12; 1988 election strategy, 209–11; and nuclear dilemma, 78; and peace movement, 1, 3–4; political divisions within, 3; rightward shift of,

Index

Democratic party (*cont.*)
  129–31. *See also* New Politics movement
Democratic Policy Commission, 129–31
Denton, Jeremiah, 105–6
DIVAD, 28, 92, 172–77
domestic programs: cuts in, 52–55
Downey, Tom, 215
Dukakis, Michael, 209–11
détente, 2, 18, 21, 28

Economic Recovery Act of 1981, 53
Eisenhower, Dwight D., 13–16, 52; attempts to limit military expenditures, 32
elections: 1980, 28–30; 1982, and freeze movement, 109–10; 1984, and freeze movement, 118–32; 1988, 209
Euro-missiles, 69; and freeze movement, 109. *See also* cruise missile; Pershing II
Exon, James, 218, 220

Fallows, James, 88–89
Fascell, Dante, 121
FBI, 106n
Fellowship of Reconciliation, 71, 73
Fitzgerald, A. Ernest, 95
Fitzhugh Commission: and procurement reform, 80–81
Fletcher, James: and SDI, 155
"flexible freeze," 213–14
"fly before you buy," 172–77
Ford, Gerald R., 22–23
Ford Foundation, 74
Ford Motor Company: and DIVAD, 173–74, 175
Forsberg, Randall, 65n, 70–71
Fossedal, Gregory, 148–49
freeze movement, 70–78; coalition-building, 58–74, 109; and conservative opposition to freeze proposal, 105–9; electoral goals of, 109–12; and 1984 election, 118–31; and nonintervention, 66–70; nuclear freeze proposal, 170; popular support for nuclear freeze proposal, 76–78; public opinion of, 106–9. *See also* peace movement
Freeze Voter 1984, 119, 123; merger with SANE, 201
Fund for Constitutional Government, 95
Future Security Strategy Study panel, 155

Gaither, H. Rowan, 14
Gaither report, 13–15, 23

Galbraith, John Kenneth, 121–22
Gang of Four, the, 69
Garn, Jake, 113
gender gap, 129
General Dynamics, 52
General Motors, 51
Gephardt, Richard, 130
Ginsberg, Benjamin, 8n
glasnost, 200
Glazer, Nathan, 22n, 129
GLCM. *See* cruise missile
Glenn, John, 122
gold plating, 84–87
Gorbachev, Mikhail, 199–200
Gore, Albert, 92, 130
Gorton, Slade, 93
Gourevitch, Peter, 7
Grace Commission, 212–13
Graham, Daniel O., 23n; and nuclear freeze movement, 148; and SDI, 140, 144–46
Gramm, Phil, 221–23
Gramm-Rudman Act, 214
Grassley, Charles, 92, 171
Greenpeace, 66n
Grumman Corporation, 215–16

Harkin, Tom, 128
Hart, Gary, 130–31; and congressional procurement reform, 171; and industrial policy, 50; and military reform, 81, 84, 89–94; and freeze proposal, 122
Hart, Peter, 124n, 127
Hatfield, Mark O., 70, 92, 110, 102–4
Helms, Jesse, 199
Heritage Foundation, 25, 95n, 145
High Frontier, 137n, 143, 144, 149
Hoffman, Fred, 155
Hollings, Ernest, 122
Hoover Institute, 25
House Armed Services Committee, 117, 219
House Committee on Government Operations: and spare parts procurement scandal, 98
House Foreign Affairs Committee: and nuclear freeze resolution, 110
House of Representatives: and nuclear freeze resolution, 110–11; and Scowcroft commission, 153
Hughes Aircraft, 51, 205
Hume, Jacquelin, 144–45

Humphrey, Gordon, 25
Huntington, Samuel P., 5

ICBM: see intercontinental ballistic missile
industrial policy: military, 49–52
infinite deterrence: contrasted with MAD, 39
Institute for Defense and Disarmament Studies, 71
intercontinental ballistic missile (ICBM): defenses for, 145; vulnerability of, 137–38
intermediate nuclear forces (INF), 198–200
interservice rivalry: and military reform, 87–88
Iran-Contra affair, 198, 200
Iranian hostage crisis, 27, 29
iron triangles, 94
Isaacs, John, 124n, 169, 201n, 202n
isolationism, 2

Jackson, Jesse, 122
Jackson, Scoop, 104, 129
Jackson-Warner resolution, 104
Jepsen, Roger, 25
Johnson, Lyndon B., 14, 15, 52
Johnston, Bennett, 143
Joint Chiefs of Staff (JCS): and military reform, 80, 87–88; and SDI, 149–50
Jones, David C., 87–88

Kassebaum, Nancy, 92, 171–72
Kaufmann, William W., 218
Kehler, Randall, 71–72
Kemp, Jack, 139, 214
Kennan, George, 23, 69
Kennedy, Edward M., 92, 102–4, 110, 221
Kennedy, John F., 14, 15–16, 58
Keynesianism. See military Keynesianism
Keyworth, George, 141, 144, 158n
KGB, 105
Kirk, Paul, 130
Kissinger, Henry, 14n, 114n
Kotz, Nick, 118n
Krings, John E. 173, 175–76

labor, organized, 200
Lamberson, Donald, 142
LaRocque, Gene, 121–22

Lawyers Alliance for Nuclear Arms Control, 66n, 74
Laxalt, Paul, 113, 140
League of Women Voters, 73, 202
Lent, Norman, 215
Levine, Mel, 171
Levitas, Elliot, 111
Lind, William S., 89n
LTV Corporation, 48–49

MacArthur Foundation, 74
McCarthy, Joseph, 12
McDonnell-Douglas, 50, 173, 204
McFarlane, Robert, 149–51, 152n, 154
McGovern, George, 61; and arms control, 122
McNamara, Robert, 69
Manatt, Charles, 122, 130
Markey, Ed, 102–4, 109n, 110
mass media, the: and the peace movement, 75–76; and procurement reform, 96–101
Maverick missile: and "fly before you buy," 172
Meese, Edwin, 144, 146, 147n, 151n, 204
Meyer, Edward C., 87
Midgetman, 117; endorsed by Democratic Policy Commission, 130
military Keynesianism, 46–52
military reform: and Congress, 1–2, 91; and conservatives, 91–92; and conventional defense, 81–91; and Democratic Policy Commission, 131; and liberals, 91–92; and 1984 Democratic convention, 123–24; origins of, 79–101; and politics of national security, 1. See also procurement reform
Military Reform Caucus, 92–96, 169–76, 171n, 179–80
military space policy, 141
minimum nuclear deterrence, 68
missile gap, 16
Mitchell, George, 92
Mobilization for Survival, 64, 66n, 67n, 72n
Mondale, Walter: and freeze proposal, 122; and 1984 campaign, 125–28
M-1 tank, 97, 172
Moscow summit, 199–200
Munich agreement: lesson of, 33–34
Mutual Assured Destruction (MAD), 33, 134, 139
MX missile, 68, 69, 112–18; and ballistic missile defense, 138; and Bush admin-

Index

MX missle (*cont.*)
istration, 216–17; and procurement reform, 169; and SANE, 64. *See also* Scowcroft commission

National Conservative Political Action Committee (NCPAC), 22
National Council of Churches: endorsement of freeze, 72
National Intelligence Estimates (NIE), 22–24
National Organization of Women. *See* NOW
National Security Act of 1947: and military reform, 80
National Security Council (NSC), 12; military space policy, review of, 141
National Security Decision Directive 13, 39
National Taxpayers Legal Fund (NTLF), 95
NATO, 65, 199
New Class: and New Politics, 60; and peace movement, 74
New Deal, 11, 31
New Frontier, 16
New Politics movement, 17, 58–62; coalition with freeze movement, 119–21; and Mondale campaign, 127; and 1984 Democratic party platform, 124; and nuclear dilemma, 78; and peace movement, 73; and political change, 203; and public interest groups, 20; rejected by Democratic party, 130
New Right, 21–22; alliance with neoconservative elite, 25–28; and Panama Canal treaty, 25; peace movement, attack on, 105
Nichols, Bill, 92, 98
Nike-Zeus, 137
Nitze, Paul, 22, 23–24, 104
Nixon, Richard M., 16, 18, 32, 80, 137
Novak, Michael, 129
NOW, 73, 120
NSC-68, 12–13, 23–24
nuclear arms control: popular support for, 76–78
nuclear weapons peace movement. *See* peace movement
Nunn, Sam, 93, 98n, 115–16, 157n, 164, 218

observation-orientation-decision-action cycles (OODA), 82–83

Office of Management and Budget (OMB), 214–15
Operational Testing and Evaluation, Office of (OT&E), 172–73
Operation Ill Wind, 204–7
operational testing, 172–77

Packard, David, 22n, 81, 206
Packard Commission, 188–94, 206, 210
Paine, Christopher, 122–23
Panama Canal treaty, 22, 25
partial test ban treaty, 58
parties, political: decline of, 18–19. *See also* dealignment; Democratic party; partisan competition; Republican party
PD-59. *See* Presidential Directive 59
"peace dividend," 221–23, 226
Peace Links, 106
peace movement, 3, 16; and coalition building, 119–21; and Democratic party, 3; ideology of, 66–78; and Mondale campaign, 127; organizational expansion of, 72–74; and politics of national security, 1; and SDI, 147–48, 201–2; support for INF treaty, 199; transformation into arms control lobby, 200–204
"Peace Plank," 123
Peace Roundtable, 123
peace through strength, 2, 66
Peace through Strength Coalition, 105
Pearl Harbor, 33, 34
Pelosi, Nancy, 213
Pentagon. *See* Defense, Department of
perestroika, 200
Perry, William, 114n
Pershing II, 65, 68, 198
Phillips, Howard, 22
Physicians for Social Responsibility, 66n, 72, 74
Pipes, Richard, 23, 129
Presidential Directive 59, 38, 65, 68
President's Commission on Strategic Forces. *See* Scowcroft commission
President's Private Sector Survey on Cost Control. *See* Grace Commission
procurement reform, 94–101; and Operation Ill Wind, 204–7. *See also* Military Reform Caucus
Project Defender, 137
Project on Military Procurement, 95–101, 174

Protestant groups: freeze movement, endorsement of, 120
Pryor, David, 92, 171

Rainbow Coalition, 122, 129, 200
Rasor, Dina, 95–97
Reagan, Ronald: candidacy of, 2; and conservative political advocacy groups, 24–25; and CPD, 28; and desirability of SDI, 147; election of, 21, 31; and "evil empire," 32–33; and freeze campaign, 105; on goals of strategic defense, 156–57; INF treaty, 199; 202; meeting with Joint Chiefs of Staff, 149; and military Keynesianism, 46–52; and Moscow summit, 199–200; and peace movement, 56, 65–66; and Reykjavik summit, 163–64; and SDI, 134–35, 139–40; similarities to Kennedy, 16; "Star Wars" speech, 51, 133, 150–51; START proposal, 104–5; and Townes Commission, 113
Reagan administration, 2–5; ABM treaty, reinterpretation of, 164; and ballistic missile defense, 138; framework of national security policy, 21; freeze movement, attacks on, 104–6; increased defense spending, 31–39, 50–51; kitchen cabinet, 145; and military reform, 79–101; and MX missile, 113–18; perception of Soviet ICBM threat, 7; polarization of nuclear arms debate, 81–82
Reagan campaign: and Democratic strategy in 1984, 124–29; and foreign policy, 28–30
Reaganomics, 2–3; and military Keynesianism, 47
Reagan revolution, 1, 11; and military policy, 31–55; and peace movement, 78
reform triangle, 94–97
Reilly, William, 214
Republican party: building the permanent military establishment, 12; and neoconservative activism, 21–22; and 1980 elections, 28–30; realignment of, 202
Reuben Fund, 95n
Reykjavik Summit, 163–64
Reynolds, Chip, 121n, 124n
Richardson, Robert, 146
Robb, Charles, 122, 130
Rockefeller Fund, 74

Root Reforms, 80
Rostow, Eugene, 22
Rowny, Edward, 32n–33n, 104

SALT I, 18, 25; and ABM treaty, 137–38
SALT II, 22, 30, 64; campaign to prevent ratification of, 26–27; and peace movement, 70
SANE, 58, 63, 66n, 73, 123; merger with Freeze, 201–2
Schlafly, Phyllis, 149
Schumer, Charles, 100
Scowcroft, Brent, 114n, 115, 216
Scowcroft commission, 114–17, 152–53
Senate Armed Services Committee: and DIVAD hearings, 174; hearings on spare parts procurement scandal, 98; interpretation of ABM treaty, 164
Senate Armed Services Preparedness subcommittee, 14
Senate Foreign Relations Committee, 112, 164
Sentinel program, 137
Shefter, Martin, 8n
Shultz, George, 199; Zablocki resolution, 104
Sierra Club, 73
Skocpol, Theda, 8
SLCM. *See* cruise missile
Smith, Denny, 91n; and DIVAD, 175; member of MRC, 92; and Military Reform Caucus, 171
Smith, Gerard, 69
Smith, Larry, 174–75
Solo, Pam, 204
Soviet Union, the 32–35; ABM effort of, 153; and NSC-68, 12–13; as perceived by Reagan administration, 7; public perception of, 108; Team B reassessment of, 23–24
Spinney, Franklin, 86–87
Sprey, Pierre, 84–86
Sputnik, 13–14
SS-20 missile, 199
Star Wars. *See* Strategic Defense Initiative
START. *See* Strategic Arms Reduction Talks
Stealth bomber, 131, 222
Stevens, Ted, 220
Stevenson, Adlai E., 58
Stimson, James A., 8n
Stockman, David, 36

Index

Strategic Arms Limitations Talks (SALT). *See* SALT I; SALT II
Strategic Arms Reduction Talks (START), 104–5, 117, 118, 198–200
Strategic Defense Initiative (SDI), 39–40, 201–2; and arms control, 163; before 1983, 141–44; bureaucratic institutionalization of, 160–63; Bush administration requests for funding, 219; corporate support for, 162–163; cuts in budget, 208; and Democratic Policy Commission, 131; early deployment of, 163; funding of, 155, 161; and Mondale campaign, 126–27; origins of, 134–54; political motivations for, 151–54; popular support for, 168; public opinion of, 146; strategic justifications for, 156–60; threat to peace movement, 131–32; transformation from initiative to program, 154–68; and White House network, 144–47
Strategic Defense Initiative Organization (SDIO), 141; and institutionalization of SDI, 160–63
Sullivan, Louis W., 215
Symington, Stuart S., 15
symmetrical containment, 32

Talbott, Strobe, 223
Team B, 23–24
Teller, Edward, 22n, 135, 144–46
Texas Instruments, 51
Thurow, Lester C., 51
Tower, John, 98n, 215
Townes Commission, 113
Trident II, 68
Truman, Harry S., 12, 13, 52

U.S. military budget, 34–38; and Bush administration, 212–23; under Carter administration, 35; drop in operating expenditures, 44–45; and "flexible freeze," 213–14; growth under Reagan, 207–8; impact on world order, 224–26; and Reagan administration, 34–38; SDI research funding, 155; and Weinberger, 34–36
U.S. military strategy; 38–44, 82–84

Union of Concerned Scientists, 66n, 72, 73

Van Cleave, William, 23
Vessey, John, 219
Vietnam, 2, 11, 16–17, 18, 21, 58–63;
Viguerie, Richard, 22, 25

Walker, Robert, 110–11
Wallace, George, 18
Wallop, Malcolm, 139–40
warfare state, 3
Warner, John, 93
Warnke, Paul, 25
War Powers Resolution, 61
Warsaw Pact, 199
Watergate, 18
Watkins, James D., 149–50, 214
Wattenberg, Ben, 129
Webster, William, 219
Weinberger, Caspar, 84, 156–58n, 165; and demands for increased military budget, 34–36; and DIVAD, 175–76; failure to comply with OT&E, 173; on military Keynesianism, 48; and "peace through strength," 33n; on Soviet superiority, 32n; on strategic defense, 141–42; and Townes Commission, 113; Zablocki resolution, 104
Weyrich, Paul, 22, 25
"whistle-blowers," 94–97
Whitehurst, G. William, 91–94, 171
Wilson, William, 144
window of vulnerability, 14, 33, 65, 115, 138, 152
Wolfe, Alan, 6n
Women's Action for Nuclear Disarmament, 120
women's groups: endorsement of freeze movement, 120
Women's International League for Peace and Freedom (WILFP), 66n; and peace movement, 120
World War II, 2

Zablocki, Clement, 102, 110
Zablocki resolution: nuclear freeze resolution, 103–4
zero option, 118

Library of Congress Cataloging-in-Publication Data
Wirls, Daniel, 1960–
   Buildup: the politics of defense in the Reagan era/Daniel
Wirls.
     p. cm.
   Includes bibliographical references and index.
   ISBN 0-8014-2442-9 (alk. paper)
   1. United States—Military policy.   2. United States—Armed
Forces—Appropriations and expenditures.   I. Title.
UA23.W486   1992
355'.0335'73—dc20   91-55075